D0264848

# SERENGETI
# A KINGDOM OF PREDATORS

# SERE

# A KINGDOM O

# NGETI

# PREDATORS

GEORGE B. SCHALLER COLLINS, ST. JAMES'S PLACE, LONDON 1973

To John S. Owen
whose vision and initiative
have helped to create in Tanzania
some of the world's finest
national parks

William Collins Sons & Co Ltd
London • Glasgow • Sydney • Auckland • Toronto • Johannesburg
First published in Great Britain 1973

ISBN 0 00 211747 9
Printed in Italy by Arnoldo Mondadori Editore
First Edition

ACKNOWLEDGMENTS: Several institutions and many individuals, too numerous to mention, made my study possible and I am grateful to them all. The National Science Foundation and the New York Zoological Society financed it and the latter organization also sponsored it as part of the program of the Institute for Research in Animal Behavior, which it runs jointly with.Rockefeller University. The National Geographic Society provided photographic assistance. To John Owen, to whom this book is dedicated, I shall always be grateful, for our years in the Serengeti were among the happiest and most rewarding of our lives. The Tanzania National Parks generously permitted me to work in the park. The Serengeti Research Institute, under the direction of Hugh F. Lamprey, provided us with facilities and much help. My wife, Kay, and sons, Eric and Mark, were so much a part of my life that without them my feelings and experiences would have been very different.

Bob Kuhn first encouraged me to do this book, and Gordon Lowther read the text critically. A fellowship from the John Simon Guggenheim Memorial Foundation and assistance from the New York Zoological Society enabled me to spend time in writing.

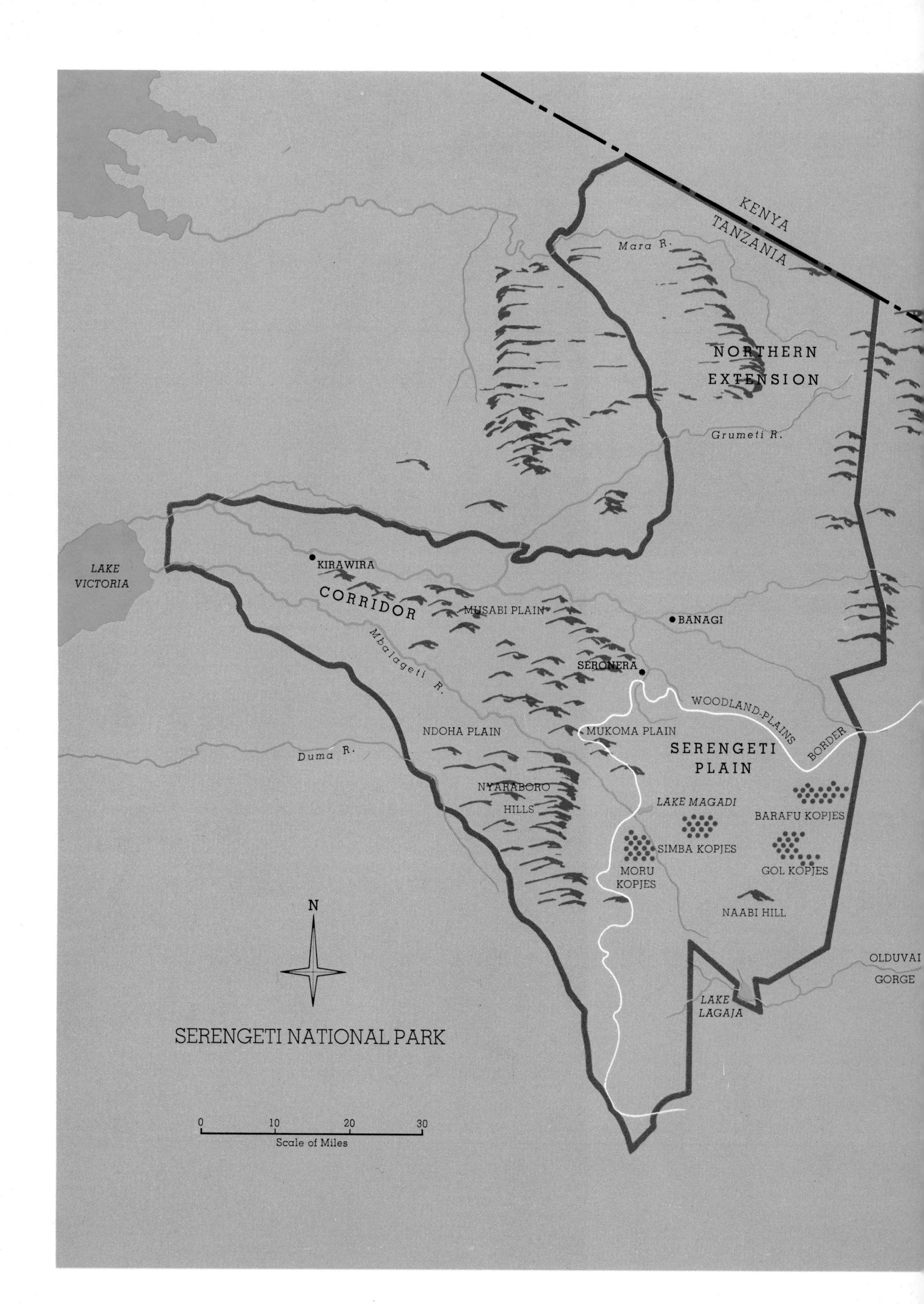
KENYA
TANZANIA
Mara R.
NORTHERN
EXTENSION
Grumeti R.
LAKE
VICTORIA
KIRAWIRA
CORRIDOR
MUSABI PLAIN
BANAGI
Mbalageti R.
SERONERA
WOODLAND-PLAINS
BORDER
NDOHA PLAIN
MUKOMA PLAIN
SERENGETI
PLAIN
Duma R.
NYARABORO
HILLS
LAKE MAGADI
BARAFU KOPJES
SIMBA KOPJES
MORU
KOPJES
GOL KOPJES
NAABI HILL
N
OLDUVAI
GORGE
LAKE
LAGAJA
SERENGETI NATIONAL PARK
0
10
20
30
Scale of Miles

# CONTENTS

# INTRODUCTION

Exalted and denigrated, admired and despised, no animals have so aroused the emotions of man as have the large predators. The lion is the King of Beasts, once venerated as an animal god; the tiger is extolled for its beauty and strength; the wolf is a symbol of the wilderness. Yet at the same time these predators have been and still are persecuted to such an extent that they have vanished from much of their former range. Sometimes such enmity can be justified, for predators may kill domestic stock and even man himself, but more often it is irrational, as if some ancient command forced man to battle an enemy. Throughout his evolution, first crouched timidly in the dark, then huddled by a fire clutching a stone or club, man has been at the mercy of a determined predator. Perhaps he now retaliates for eons of harassment, subconsciously evening the score. But there may be a further explanation. Though by inheritance a vegetarian primate, man has been a predator, a killer of animals, for at least two million years. His interests and emotions, his whole outlook on life, are those of a hunter and as such he views other predators as competitors for space and food. Just as a lion tries to kill a leopard he encounters, so man attacks a wolf, relieving his primordial apprehensions. Still, while man fears the predators, he secretly exults in their power, he feels a contagion, an emotional kinship to them. It is no coincidence that visitors to the African parks watch not the impala and zebra, but the lion and leopard. Even in sleep these big cats convey a feeling of barely contained strength, an ever-present threat of death, which man the hunter finds satisfying, though the danger is vicarious from the safety of a car. Our dual past still haunts us. We hear a lion roar and the primate in us shivers; we see huge herds of game and the predator in us is delighted, as if our existence still depended on their presence.

The large predators have now reached the twilight of their existence. Shot for so-called sport, in demand by the fur industry, trapped and poisoned because they kill both wild and domestic animals in which man has a vested interest, the tiger, cheetah, wolf, and others will probably not survive except in large reserves. In the future only national parks may remain as samples of wilderness where man can renew his ancient ties with the predators that were once his competitors and with the prey that provided him with sustenance.

To preserve a species adequately, especially one that is confined to a small area, it is necessary to know how far it travels, how often and how much it eats, how fast it reproduces, and so forth. Research on a predator is particularly complex, for it is essential to study not only its biology but also that of the various animals on which it preys. A simple question such as "What effect does predation have on the prey populations?" requires years of work before even a tentative answer can be given. Throughout the world, including North America, predators are being exterminated on the assumption that they harm wildlife. The relations between predator and prey are often thought to consist of an elementary equation—lion + zebra = lion. Of course, to see a leopard rush from concealment and in a flurry of violence pull down a gazelle or to watch a wolf pack harry a moose until it sinks onto the bloody snow is to view an event of such unrestrained drama that it leaves the average observer not only shaken but also outraged. There is undeniably one fewer animal. But the fact of predation must not be confused with its over-all effect on the prey; it is essential to think in terms not of the individual but of the population. Was the killed animal sick or old and doomed to die soon anyway? Was it a surplus individual whose loss had no effect on the population? Only research in a particular area can determine in what way the predators affect their prey, yet few such studies have been made. For years the United States government has been waging a war of extermination against the puma, yet it was not until 1964 that an intensive study of the animal was begun in Idaho by Maurice Hornocker. After observing these cats for several years in one area he concluded "that elk and deer populations were limited by the winter food supply, and that predation ...was inconsequential in determining ultimate numbers." Working on wolves on Isle Royale, a small island

in Lake Superior, David Mech noted that these predators particularly took young moose, old ones, and those physically below par, and he felt that the wolves kept the moose herd healthy by preventing its increase to a level where it might damage the habitat through overbrowsing.

But facts such as these have so far done little to change opinion in favor of predators. Attacking a person's unreasonable feelings about these animals is a little like questioning his religious beliefs: he becomes ever more violent in defense of his faith when there is reason to doubt it. And so the predators continue to decline.

John Owen, former director of the Tanzania National Parks, asked me to study lion predation in the Serengeti National Park. Whereas in North America there is usually but one large predator preying on one or two prey species, in the Serengeti five major predators have over twenty kinds of prey available to them. To explore the intricacies of this complex animal community intrigued me and I was delighted to take on the task. In June 1966, my family and I moved to Seronera, the park headquarters, and remained there for a little over three years. We lived in a pleasant bungalow shaded by a flat-topped acacia tree. Giraffe sometimes browsed on this tree, lions padded by at night, and at certain seasons thousands of zebra and wildebeest thundered past on their migration. Each morning, with the first light, I searched for the big cats and observed one kind or another, and on many nights too I followed them through the darkness in my Land Rover. I recorded their movements, their social contacts, and their food habits; I censused prey; and, in general, I collected the many small facts which I hoped would contribute to an understanding of predation in the Serengeti. Sometimes Kay or my sons, Eric and Mark, accompanied me, but usually I worked alone, my daily schedule patterned after that of the animals I had come to study.

On the following pages I present some of my findings; the more detailed substance can be found in my scientific report *The Serengeti Lion,* published by the University of Chicago Press. Most of my work was devoted to the lion, a beast that more than any other adds authenticity to the African scene, and much of the book is therefore devoted to it. But the lion is only one part of a whole, and to understand its role in the Serengeti community, the other large predators as well as their prey must also be considered.

As a scientist I am by tradition expected to control my experience rather than yield to it. In this I failed yet am not displeased to have done so. I cherished my escape from the organized lunacy of life in the city to the elemental complexity of the wilderness. Yesterday, today, and tomorrow became one as I lived for the moment only, seeking a sense of unity with the earth, the animals. As weeks and months passed, I learned to recognize many of the lions and other predators. Ceasing to be mere animals, they became individuals about whose problems I worried and whose future I anticipated. They became part of my memories of an austere and seemingly harsh land in which man seems of no consequence until he imperceptibly becomes a part of it. This life of isolation, of spending hours alone each day with animals, was part reality, part illusion, and it suited me; it was neither a denial of life nor a retreat from those I loved, but a way of sustaining a sense of spiritual independence. Solitude provokes reflection, and the study became a quest for understanding, not just of the predators but also of myself, a personal *corrida.*

The most pleasant way to approach the Serengeti is from the east, across the highlands, tracing the rim of Ngorongoro Crater, where wisps of cloud finger through the mountain forests, then dipping down into the grassy Malanja depression before suddenly topping a rise to see the plains far below, stretching boundlessly to the horizon. Gently rolling, they extend for some two thousand square miles. "And all this a sea of grass, grass, grass, grass, and grass. One looks around and sees only grass and sky," exulted Fritz Jaeger, who in 1907 was one of the first Europeans to visit the area. It is an exciting view, though alien in its immensity; the eyes roam, seeking a place to rest, but the plains seem without end except for the Gol Mountains in the north, subdued and featureless in the sun. The tortuous road drops down the escarpment and then heads west, at first flanked by an occasional acacia, then cut by the wooded depression of Olduvai Gorge, whose depths have revealed

more of man's history than any other place on earth, before finally reaching the open plains. Trailing a ribbon of dust, one drives on without landmark, the horizon without depth and perspective, until at last Naabi Hill rises ahead, humped like a stranded whale. Here in the eastern plains the grass is short, but toward the west, where the soil is deeper and there is more rain, it reaches a height of three feet, measureless waves of stems which no plow has ever touched.

Jumbled piles of granite, the kopjes, jut from the plains like a string of islands, the boulders polished smooth by the wind and rain. These kopjes are worlds of their own. Fleshy-leafed aloes grow there and in moist clefts sometimes a *Gloriosa* lily, its crimson blossoms incongruous in this landscape of modest tones. I liked climbing in the kopjes, though always hoping that I would not disturb a lion or leopard in its shady retreat. Occasionally one finds some paintings, of shields and animals and little stick men, done with charcoal and ochre by the pastoral Masai, who had ruled this area for over one hundred years before it was made into a national park. The work is crude and recent, yet I took pleasure in finding this hidden art, rarely seen now except by the red-and-blue *Agama* lizards that scurry over the rock faces. As one sits on a kopje, eyes squinting against the brilliant light, the plains distorted by quivering heat waves, the imagination conjures up visions, visions that inevitably dissolve into mere kopjes and zebra.

Seronera, the park headquarters, lies at the edge of the plains among kopjes and widely spaced acacia trees, whose feathery leaves diffuse the harsh rays of the sun. To the north and west and south spread sparse woodlands, a parkland gray-green in hue, subdued as if the light has leached its richness. There are many kinds of acacia—flat-topped umbrella acacia, which to many are the quintessence of Africa; fever trees with lemon-colored bark; thickets of whistling thorn with *Crematogaster* ants scurrying in and out of the galls in which they raise their young. *Commiphora* trees are there too, gnarled like apple trees. The woodlands spread across valleys and plateaus and over hills that rise a thousand feet above the level of the plains, particularly in the western part of the park. The terrain slopes gently and irregularly from an altitude of about six thousand feet in the eastern section to about four thousand feet near the shores of Lake Victoria. The streams flow westward too, the Grumeti, Mbalageti, Duma, Bologonja, and Mara, to name just a few, but only the last two are perennial, whereas the others contain pools during the dry seasons of the year. It is a vast, austere land, beautiful though without gentle charm, mostly lacking the comfortable views that give a person confidence and linger in his memory. Only along the banks of streams, perhaps, in the shade of a fig tree where the air is suffused with the heavy odor of ripe fruit and where nearby a fish eagle may fling its wild scream to the sky, is there refuge from the sense of isolation and exposure.

The whole tempo of life is governed by the rains. The dry season begins in May and stretches into November. First to dry, the plains soon present a bleak appearance, their prevailing color yellow and brown. Dry stubble crackles underfoot and dust devils twirl over the rises. With the grasses shriveled by the sun and water confined to a few alkali pools, most hoofed animals trek off the plains in May and June, zebra and wildebeest first, followed by Thomson's gazelle, heading for the woodlands, where food and water remain. Few animals stay behind: a forlorn jackal nosing under a desiccated wildebeest dropping in search of dung beetles, ungainly ostriches swaying along until swallowed by the haze, an occasional Grant's gazelle.

After the tall grasses have gone to seed and turned yellow, fires sweep across the park, removing undergrowth, killing saplings, scorching tortoises, destroying ostrich nests. Flames assault thickets, reducing them in size, then devour dead trees, leaving them as ashy skeletons on the ground. Soon the land is black and bare, the rocky slopes exposed to the elements. Little is left for the grazers to eat and it is a lean time of the year for them, a time when the weak may die of malnutrition and disease. Man sets these fires both inside and outside the park, sometimes to

stimulate a fresh growth of green for his livestock but more often through carelessness or from a mere pyromaniacal impulse. Paradoxically, the fires both harm and benefit the grazers. Antelopes have had their major evolutionary development in grasslands and many are adapted to it in tooth structure and digestive system and in other ways. Without fires, large portions of the Serengeti would ultimately be covered with a dense tangle of brush, which in turn would reduce the number of plains animals. However, at present elephants are uprooting trees and annual fires prevent widespread regeneration, saplings needing at least three years of immunity from fire to become established, with the result that the woodlands are retreating and the ecological diversity is threatened.

Flat-bottomed white clouds drifting before the wind like sailing ships herald the onset of the rains. Then towering thunderheads balanced on black pillars of rain bring the first local showers. Almost overnight new blades of grass push through the crusted soil and soon the land radiates a brilliant sheen of green. *Ramphicarpa* blossoms litter the plains like scattered pieces of white tissue paper. Birds begin to court feverishly. A pair of crowned cranes dance with graceful leaps, a male bishop bird flies upward from his reedy perch like a ruby tossed into the wind. Kori bustards patrol the high ground on the plains, flashing their white inflated throat sac and undertail coverts like heliographs from hilltop to hilltop, proclaiming their territory. At this season too European storks, winter visitors from Russia and Poland, wheel in flocks, mere black specks high in the sky until, in turning, their bodies gleam white in the sunshine. With the grass greening on the plains, the wildebeest, zebra, and Thomson's gazelle return. Why do they surge back at the first opportunity, leaving behind the good forage in the woodlands? Perhaps the short grasses are particularly nutritious, perhaps it is a matter of tradition, that for eons this is where the plains animals have come, a pattern established long ago when the area may have been more densely wooded.

The rains in November and December are erratic and fail in some years, thereby forcing the herds to stay in the woodlands or to eddy back and forth unpredictably. But usually by the end of December the grasses in the woodlands have reached their full height, whereas the central and eastern plains look like a well-kept lawn, trimmed by thousands of mouths. January and February tend to be fairly dry. Torrential rains may fall in March and April, with at least a third of the average annual rainfall of thirty inches soaking the land at this time. The rivers become swollen, the roads often impassable; lakes Lagaja and Magadi fill their alkaline beds and once again lesser flamingos impart a rosy hue to the shallows. Then, sometime in May, the skies remain clear, the dry season begins.

THE HUN

One can view the Serengeti in several ways: the pleasing rapport between terrain and vegetation; the weather, the incessant winds, the last streak of evening gold on a shadowy hill. But above all the Serengeti represents animals. It is an area throbbing with an inexhaustible vigor of life, a Pleistocene vision of immense herds, the largest such concentration in the world. Always moving, the animals spill over the park's borders, even its five thousand square miles unable to contain them, for tradition dictates that some of the migratory species use ten thousand square miles. To name just a few of the most abundant species, there are now over 500,000 wildebeest, 180,000 Thomson's gazelle, 150,000 zebra, 65,000 impala, 50,000 buffalo, and 25,000 topi. One is almost never out of sight of animals, and the view described by Stewart White in 1913 in what is now the northern part of the park still exists today: "Never have I seen anything like that game. It covered every hill, standing in the openings, strolling in and out among groves, feeding on the bottom lands, singly, or in little groups. It did not matter in what direction I looked, there it was; as abundant one place as another."

On the plains the wildebeest are pervasive. Sometimes I walked among them after thousands had concentrated on a green flush. Retreating a few hundred feet to let me pass, the animals nearest me stand mutely, the sun glinting on their horns, those farther away grunting like a chorus of gigantic frogs. The air is heavy with an odor of earth and manure and the scent of trampled grass. During the westward migration the wildebeest often move in long lines, their white beards gleaming. They lope along with a curious hunched gait, pouring over the rises, funneling down valleys. If a river bars their way, they plunge in, those that hesitate being swept on by the mindless horde behind, until the water is crowded with bellowing animals thrashing to get up the slippery banks. Finally the herd has crossed and rushes on, and sometimes several bodies turning silently in the eddies attest to its passing. As the animals sweep westward the huge herds break up and move in small groups northward, where near the Kenya border they tarry until once again the rains come to signal their return to the plains, their whole existence revolving around a search for food to sustain their endless masses. 

Many zebra also migrate, though they travel in smaller herds than do the wildebeest and their pattern of movement is more erratic. Herds are not mere amorphous aggregations but divided into distinct family units, consisting of a stallion, several mares, and their young, which may remain together for months and years. Zebra often migrate first through an area, eating the coarse grasses which other species disdain. They stream through, braying wildly, and as one watches their black-and-white columns in the flat light one becomes dizzy from the motion until only the zebra seem stable in a moving world.

Thomson's gazelle do not migrate as far as the other two species, many of them remaining around Seronera along the edge of the plains during the dry season. Most other species are either semimigratory, shifting their range with the available forage, or sedentary. The latter vary from the diminutive dik-dik, a pair of which may spend its life on an acre of ground, to a buffalo herd, whose range may encompass one hundred square miles of terrain. Some twenty-five species of hoofed animals and the elephant inhabit the Serengeti, a variety possible only because each has adapted to a particular niche. Each species harvests only a certain portion of the resources, especially in times when supply is short: warthog root, gazelle crop tender blades of grass and buffalo coarse ones, rhinoceros browse low on shrubs, giraffe on tops of trees. In spite of this intricate ecological division of labor in harvesting resources some niches are possibly unoccupied. Today's fauna is impoverished. During the Pleistocene, about 30 per cent more genera existed than now, including an antlered giraffe (*Libytherium*), a horse (*Stylohipparion*), a large bovid named *Pelorovis,* and the *Dinotherium* elephant. Gone too are several predators, notably a giant hyena and the saber-toothed cats. With what wonder and excitement would a zoologist transported backward in time view such a primeval scene.

The various prey species, each so different in appearance and habits, have one main point in common: they have endured. They have achieved a tenuous balance between the opposing forces of birth and survival on the one hand and death through disease, starvation, and predation on the other. A constant evolutionary race exists between prey and predator. As prey develop some slight advantage a predator usually finds a way to overcome it—yet the race can have no winner. The hunting methods of predators have evolved around two basic types, the stalkers, which attempt to approach undetected, and the coursers, which have the speed and stamina to pursue in the open. Encountering prey, I almost subconsciously evaluated the possibility of catching it; I identified myself with the predators. By doing this, and by watching predators hunt, I soon realized how much the appearance and behavior of prey have been influenced by constant predator pressure. 

Without predators, prey would have no need to run at forty miles per hour, it would not require a sense of smell powerful enough to detect danger at several hundred feet or bulging eyes for wide-angle vision, and it would not always have to remain alert, seldom sleeping for more than a few minutes. The zigzag run of a gazelle, the twisting leap of an impala, the bunching-up of zebra, all are traits that make it difficult for predators to select and pursue an individual. Births proceed quickly. At no time in its life is an animal more vulnerable than right after birth. But in five minutes a wildebeest calf struggles to its feet, its powerful racial wisdom urging it to follow its mother, remaining by her flank at all costs, because to become separated in the vast herds means certain death from starvation unless the calf is rescued from this fate by a predator. Any behavior which reduces meetings between predator and prey benefits the latter whether or not it evolved for that purpose. Wildebeest have an annual birth peak timed in such a way that there is ample nutritious forage for the mother and young. With thousands of newborn available, predators can lead a leisurely life, gorging until they can hold no more, but after a month or two the youngsters cease to be highly vulnerable. Fewer die in this way than if births were evenly distributed throughout the year.

Gazelle young spend the first week or two after birth crouched motionless and alone while their mother grazes some distance away. Predators often overlook such concealed animals. Adult reedbuck also behave in this cryptic fashion, and I have seen lionesses rest within twenty feet of such an animal without sensing it. When discovered, a reedbuck explodes so suddenly from its grassy retreat that a lion more often than not just stands there with a bewildered look on its face. But only small and fairly solitary species inhabiting dense vegetation can hope to escape in this way. Another alternative is to form herds. The eyes and noses of many animals are better at detecting danger than those of just one. Usually at least one member of a herd has its head raised, scanning, sniffing, while the rest eat.

Topi males stand on termite hills, thereby advertising their territory, but at the same time they can survey their surroundings. Zebra often rest their heads on each other's backs, a friendly gesture which also gives them a 360-degree view. Traveling in single file, herds decrease the probability of stumbling on a hidden predator. Warthog retreat into a burrow at night. Most hoofed animals, including even the normally silent giraffe, snort in alarm when they sense danger, a warning system that all understand. Some species such as buffalo and giraffe have grown so large that only lion have much success in killing them, and rhinoceros and elephant have almost escaped the effects of predation. Selection has helped prey to survive in an impressive number of ways.

Each individual must also learn the habits of each predator and the degree of danger it represents. For instance, Thomson's gazelle almost ignore jackals, whereas they view a hyena warily and a lion or leopard is usually watched from at least one hundred feet away. Cheetah sprint fast but not far and gazelle prefer to have a safety margin of about three hundred feet. But hunting dogs, which can run at forty miles per hour for several miles, are avoided as soon as a pack moves into view. Each youngster must learn from the adults how closely it may approach a predator with impunity; it finds out that a visible lion is a safe lion, that waterholes should be visited in daytime, when it can best spot danger. Newborn animals lack such knowledge; they have no fear of predators. Indeed, I once saw a lost wildebeest calf trustingly plod behind two male lions for over a mile, drawn by its innate tendency to follow something, anything.

Many prey animals carry weapons which they may wield in defense. Zebra may slash with the hoofs of their hindlegs at hyenas, giraffe may flail with their front ones at lions. Although horns developed for combat between members of the same species, they may also be used against predators and many an inept lion has been impaled on the formidable horns of buffalo. Zoologist John Goddard related an incident he observed in Ngorongoro Crater. A young male lion harried a rhinoceros calf. The mother of the calf "wheeled around with incredible speed and gored him twice in the centre of the ribs, using the anterior horn with quick stabbing thrusts. The lion rolled over, completely winded. The rhinoceros then gored the lion once in the centre of the neck, followed by another thrust through the base of the mandible, killing him instantly." Such defense depends on the size relation between predator and prey: a gazelle may butt a jackal but not a cheetah, a wildebeest may attack a cheetah but not a lion.

Of course the behavior of an animal is finely balanced between its need to avoid a predator and its need to fulfill other requirements. Zebra must drink every day or two even if their visit to a waterhole results in the death of a family member; a wildebeest bull must defend a territory and ceaselessly herd his harem of cows even if a lion takes occasional advantage of his preoccupation. These are calculated risks, but the price of actual carelessness sooner or later is death. I watched one zebra foal sleep soundly until a lion abruptly wakened it. Another time a warthog sauntered to a waterhole past several waiting and obviously nervous gazelle. A lion made certain that such impetuousness was not repeated. Selection tends to be for conformity, and a gazelle that shows its individuality by veering from a bunched herd to seek its own route of escape will be chosen by a pursuing cheetah or hunting dog. On the other hand, an animal that gives in to the mindless mass of the herd is in danger too. Migrating wildebeest may surge through thickets and across streams disregarding the presence of lions, drawn as if by a collective will. A member is killed, the herd mills briefly, then plunges on. One day several lionesses captured six successive wildebeest at the same river crossing, neither the bodies nor the smell of blood halting the flood of animals.

No animal can remain constantly alert or avoid all dangerous situations. Prey is not always fearful. Evasive action is taken when a predator is sensed, but otherwise the animal leads a mundane life, consisting of feeding, resting, moving, mating, fighting. Even though a member of a herd is captured, the others merely look and then return to their former activity, the incident closed.



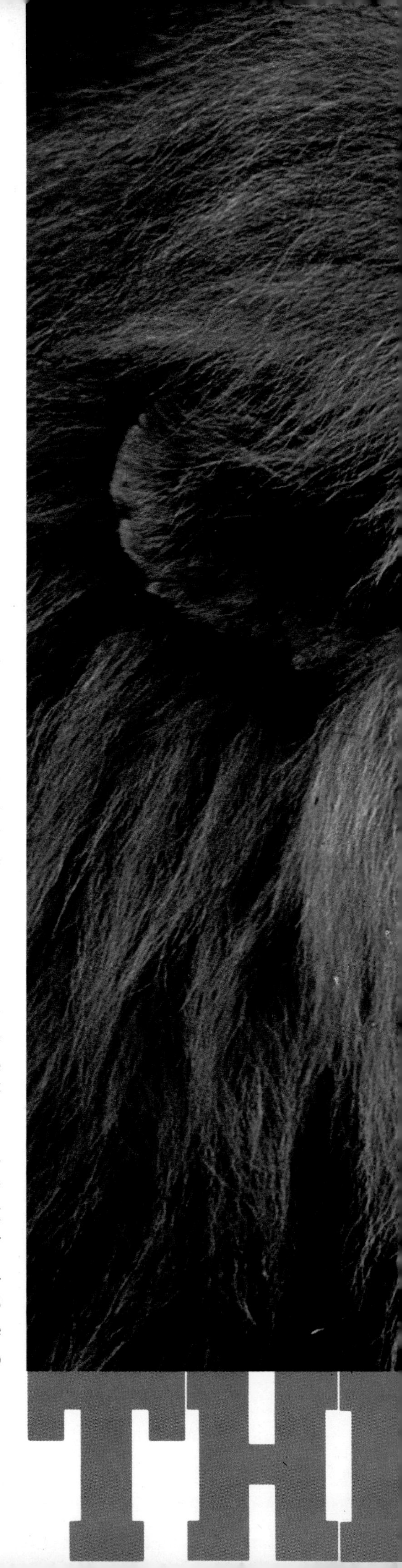

# HUNTERS

For some two million years man has continuously been part of the hunting scene in the Serengeti. At first, like any predator, he killed to the best of his ability, though perhaps using simple weapons to club and thrust. Yet he evolved in and remained a member of the ecological community, controlled by the forces that affect all living things. Frail and slow of foot, a primate seeking to be a predator, man could hope to succeed only by being adaptable. More than any other animal he followed an evolutionary path of learning, rather than depending on the slow process of genetic change, and he then transmitted this information culturally. At some stage he began to think novel thoughts, he achieved a self-awareness, he added a future to his existence, and he began to shape his destiny, striving to fulfill his dreams. With that giant and unique step, he placed himself outside the natural community and became a conqueror with intent to ravish and destroy. Such is the position of man the hunter in the Serengeti

today. Where once skill was required to ambush a zebra, now even a child can set a snare, killing unselectively any animal that passes down the lethal trail. Where once hunters killed only for their needs, they now slaughter for the market in organized expeditions that include men who set the snares, others who butcher the carcasses, and porters who carry the bloody loads to town under the cover of darkness.

Once I accompanied Warden Myles Turner on an anti-poaching patrol along the southern park boundary. We saw little wildlife and much of that was shy. Following the edge of a dry river in the Land Rover, Okech, the leader of the patrol, signaled a halt along a stretch where the brush had escaped the dry-season fires. Fresh human footprints traced the sandy bottom of the stream bed. With two guards walking down one side of the stream and two down the other we proceeded on foot. Suddenly three men flushed ahead, and bounded through high grass, only their heads and shoulders visible. The guards pursued, obviously delighted with the chase. Dodging across the river in an

attempt to elude one set of guards, the poachers ran into the others, trapped by a cooperative hunting method not unlike that used by lions. Whimpering, they rolled onto their backs in submission, glanced at the rifle butts poised above them, then docilely permitted themselves to be handcuffed, a preview of the ultimate struggle, not man against nature but man against man. Each poacher carried a bow and a quiver of poisoned arrows. Their camp yielded only eight snares and two butchered carcasses, but another camp, a big one with many thick slabs of meat hung up to dry, contained the following animals, the result of only a few days' poaching: 12 Thomson's gazelle, 6 zebra, 6 warthog, 5 impala, 5 giraffe, 3 wildebeest, 2 buffalo, 2 kongoni, 1 roan, 1 eland, 1 topi, 1 reedbuck, 1 lioness, and 1 white-backed vulture. The larder of a super-predator, who, if allowed to practice without restraint, would soon destroy the last of man's primeval habitat and with it the roots of his nature.

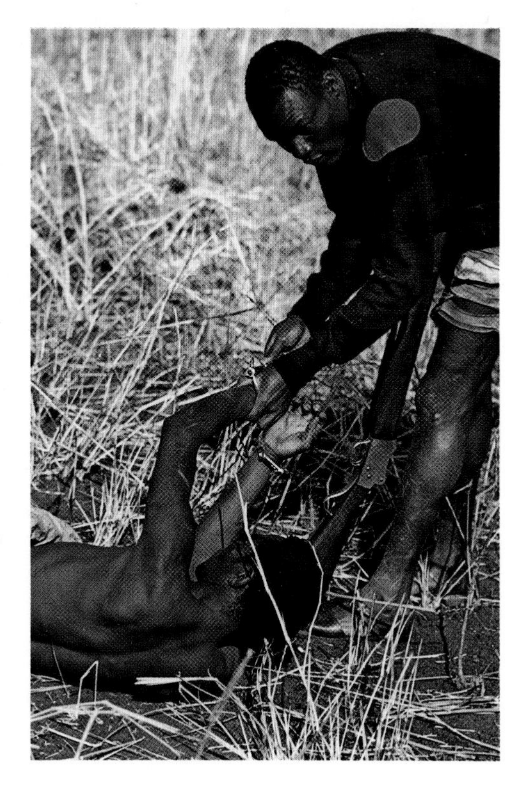

But I had come to study neither man's predation nor that of the many small predators which are part of the community: the bat-eared fox peering bright-eyed from its den, the martial eagle tearing with hooked bill at a gazelle fawn. Nor did I come to observe jackals. There are two common species in the Serengeti, the golden jackal of the short-grass plains, and the black-backed one which prefers the woodlands. However, both may congregate around the same lion or hyena kill, busily trotting around, darting in to snatch a morsel, always nervously in motion. After the other predators have left they must compete with the scavenging birds for scraps and they do so by leaping in futile fury at the vultures and marabou storks crowded around. Usually a pair of jackals has the opportunity to scavenge only when something has died or been killed in their small territory. At other times they must hunt for themselves, anything from beetles, lizards, and mice to Thomson's gazelle fawns.

Much of this book is about death, but a death so that others may live, death out of need not amusement, impersonal. Just as each prey species has evolved means for escaping predators, so each predator has evolved specialized means for capturing prey. A predator's whole existence revolves around the quest for food, and its structure, its social system, its mentality have all been modified to this end. This does not imply that the various species behave the same or even similarly. Lions are territorial, hunting dogs not; lions live in a patriarchy, hyenas in a matriarchy, hunting dogs in almost absolute equality; some species are social, others solitary. To discover these differences and to speculate about their selective advantages were among the most fascinating aspects of the study. But the Serengeti predators have one thing in common: they spend some twenty hours a day doing nothing, and when they finally rouse themselves it is usually for a meal. Hence, hunting and killing take up a large amount of space in my notes. I must admit, too, that I enjoyed watching most hunts, for at no other time do animals convey such a high level of mental and physical tension, a struggle of life and death at its most elemental.

# HUNTING

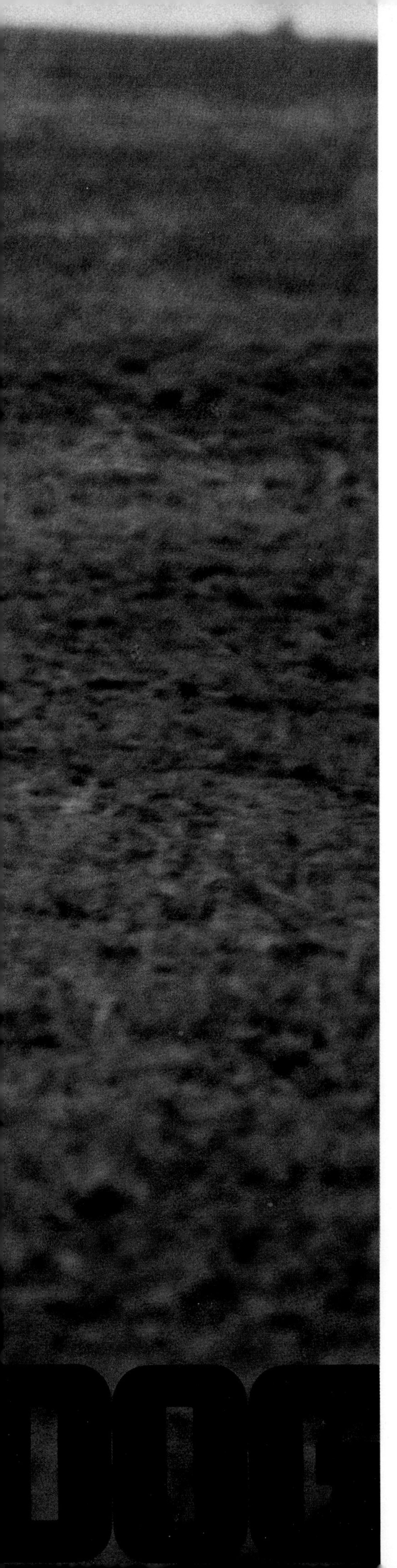

When man characterizes animals he is more often wrong than right: the owl is not wise, the dove not gentle. And the hunting dog is not the ferocious, lustful killer of popular accounts but a rather unimpressive, gentle creature with a social life that evokes a special responsiveness in us because we share many ideals. Even though hunting dogs have been so persecuted that they are now rare or absent in much of their former range throughout Africa, they retain a fatal curiosity and lack of fear of man. For instance, while I was following a pack on a rainy morning, one wheel of my car dropped into a warthog burrow. As I inspected the car, the dogs inspected me, standing in a semicircle forty feet away, their huge ears cocked, their heads turned quizzically to one side. When they continued the hunt in their tireless trot, I followed on foot, keeping at least three hundred feet behind them until after a mile they killed a zebra.

Hunting dogs live in packs of about ten animals each, but some numbering over forty have been recorded. Ceaselessly on the move and roaming widely, packs spend a few days in an area until a mysterious urge drives them on regardless of the amount of prey that remains. One pack left Seronera only to appear near Olduvai Gorge some forty-five miles away. In part this peripatetic existence is under the influence of the pack leaders, of which there may be one or several, who initiate movements, choose direction of travel, and select the quarry during a hunt. For several days I had been observing a pack consisting of ten males and one female—an unusual composition, for in most packs the sexes are represented more equitably. Leading the hunt one morning was a rather scruffy male. He passed among thousands of gazelle and other prey, drawn by some elusive vision, until he spotted several zebra after sixteen miles. These he attacked. The stallion defended his family vigorously, kicking and biting, but after a concerted effort and with much difficulty the pack dispatched a foal. The dogs then returned to the den where they had left their pups, an arduous day, yet none balked at the quixotic gesture of the leader.

Once a year, most often in March or April, a pack selects a den, often an abandoned hyena warren, in which to raise its pups. By then the pregnant female, or on occasion two females, is grotesquely swollen, a not surprising condition, for soon afterward each may give birth to as many as sixteen pups. These remain underground until, at the age of about three weeks, they begin to bumble and yip around the den entrance, their white-tipped tails raised. The pack members seem to rejoice in their presence and compete so exuberantly in nuzzling, licking, and even rolling on them that the pups may retreat to the peace of their burrow.

Once or twice a day, in the early morning and late afternoon, the pack prepares to hunt. First the animals greet each other by pushing their noses into each other's muzzles; they dash around in play, leaping against and over one another, their tails wagging, synchronizing their mood much as man does in a pep rally before a communal endeavor. One or more dogs stay with the pups, acting as guards, while the others move out, searching the horizon for prey. There is something relentless about their smoothly trotting forms, an atmosphere of impending violence as if the death of some individual is preordained. Indeed, in almost 90 per cent of their hunts they capture at least one animal, and, having learned this, a hyena or two may tag along, accepting the risk of having their rumps nipped in return for some scraps.

Hunting dogs have been castigated for their brutal way of killing. To be sure their method is untidy, but a small prey is torn apart so quickly that its death throes are brief compared to those accompanying the neat but slow strangulation usually practiced by the big cats. Only when dogs kill a large animal, such as an adult wildebeest, does the process of killing last a long time, as much as an hour. Nature has neither cruelty nor compassion. The ethics of man are irrelevant to the world of other animals. Dogs kill out of necessity, in innocence not in anger, hardly a situation to engender revulsion on the part of man. Dogs are also said to be wanton killers, and it is true that at times they may kill more than they can eat. For instance, one pack provided itself with about twelve pounds of meat per dog per day at a time when wildebeest calves were abundant, far more than it required. But the excess is not wasted, for it feeds hyenas and jackals, cutting down on their need to kill. And periods when prey can be easily caught are rare indeed in the life of a predator.

Crowding around a carcass, the dogs bolt their food, each sharing in the spoils without dissension, without the growling or snapping that often takes place among domestic dogs. When two hunting dogs want the same bone, one may give it up only to casually retrieve it later, or both may lower their forequarters in submission. With both animals behaving amicably a fight is not possible. One time I saw a dog with a withered leg arrive tardily at the kill after all meat had been eaten. He begged from the others and some meat was regurgitated for him. After gorging themselves the hunters head back, carrying the meat for the rest of the pack not in their mouths but safely stored in their stomachs. Tails wagging, the guards greet them near the den and beg. They receive a portion of meat and then so do the pups. 

A wildebeest herd is sighted. The dogs lope closer, then suddenly bunch up and advance at an ominous walk like a gang of street toughs before a fight. At first the wildebeest stand and look, then wheel and gallop off. There being no calf in the herd, the dogs merely watch its retreat. Farther on is a second herd and in it a small calf, conspicuous in its light brown natal coat. The dogs' tails whip up and they dash ahead, soon overtaking the herd, their intended victim now in the center of the massed animals. Some dogs follow the wildebeest, others run up beside them and even ahead of them, dodging in and out, attempting to scatter the herd; the air vibrates with the bleats of wildebeest and the yips of excited dogs. Finally, after a chase of less than a mile, the calf lags. At first its mother tries to place herself between it and the dogs, but a few nips at her legs send her racing ahead, abandoning her offspring to its fate. Pulling the calf down in a cooperative frenzy, the dogs quickly tear it open, bolting the viscera, then the meat, and within minutes little except skin and bones remain.

Both males and females share the tasks of hunting, guarding, and feeding the young, an equality sought so far with only limited success in human society. In fact the males are so competent and devoted to the pups that the help of a female is not essential after the first few weeks. A fascinating incident was reported in which the only female in a pack died while her pups were still small, yet the males raised them successfully. A division of labor, with some hunting while others guard, cooperative pursuit and killing, and an equal sharing of resources no doubt featured importantly in the evolution of human society. It is noteworthy that these habits are found not among monkeys and apes, but in the hunting dog, an indication that some of the same selective forces may have operated in shaping the social systems of both social carnivores and man.

By the age of about eight weeks the pups have almost lost their baby features. Their muzzles and legs have lengthened and their gray-and-white natal coat has been replaced by the haphazardly splotched one of adults. Occasionally they try to follow

the adults when these leave on a hunt. At first they are led back to the den, but they persist and suddenly, when they are about ten weeks old, the adults decide to abandon their home and return to a nomadic existence. The special treatment of pups in dog society is revealed better now than at any other time. I remember my first observation of a hunt. A pack of nine was in pursuit of a Thomson's gazelle, a favorite prey, with two dogs streaking close behind the quarry while the rest trailed. But, when the gazelle began to veer to one side, the laggards took a shortcut across the arc of the circle and surrounded the hapless animal. The dogs had just taken a first bite when their three pups topped a rise, hurrying after the pack as fast as their short legs would carry them. The hungry adults drew back from the kill. And the pups fed alone until satiated while the others waited, a sublime sight deeply appealing to our human ethics. Indeed, this behavior is a form of racial morality in that through such inborn behavior the pups are assured a meal and the circle of adults protect the kill from marauding hyenas. Pups retain this priority until they are about eight months old.

But when nature takes special care of something there must be a good reason. Hunting dogs are highly efficient predators with a seemingly unlimited supply of food in the Serengeti and they have a tremendous reproductive potential, yet they are scarce in the park and have been so for many years. Some females fail to have litters as expected and others mysteriously lose theirs. One night I slept beside a den to find out what dogs do during the hours of darkness. A violent rain suddenly lashed across the plains driven by a howling wind that made my Land Rover tremble. While most dogs huddled on bits of high ground, a female dragged her fourteen pups one by one out of the flooded den. Soon the dark clouds parted to reveal the moon. With its silvery light reflecting on the water, the plains shone like an arctic waste, and whimpering through the night, piled on each other for warmth, were the soggy pups. By the following evening five were dead.

A more tragic incident gave me a clue to the fate of many dogs. In March 1968, Hans Kruuk told me of a den he had found on the plains. There were ten adults and sixteen pups, and all were still in fine condition when I last saw them in May. In August the pack showed up near Seronera, thirty miles west of its den. Two pups were missing and several animals were obviously ill, with emaciated hindquarters and a grayish mucus that drained from their eyes. Eight dogs died within a week and still the disease raged on. One evening I caught a staggering pup to take it to a veterinarian for diagnosis. An autopsy showed canine distemper. Of the twenty-six dogs only five adults and five pups survived. The following year the pack raised seven pups. But when would the disease strike again? The hunting dog is one of nature's underdogs. Instead of ruling the predatory kingdom with their prowess, they appear to have been brutally penalized for it, and, with a helping hand from man, the species seems to be sliding slowly backward toward the dark abyss of extinction.

YENA

If disturbed at the earthy hollow where it whiles away the daylight hours, a spotted hyena is a most unprepossessing creature. It slouches off, tail tightly tucked between its legs, a skulking form with sloping back, looking over its shoulders with a harried look. Though not elegant, its body design is marvelously functional. There may be poetry in the flowing movements of a leopard but only a hyena can run at over forty miles per hour, keep up a steady lope for miles, and gain sustenance from bones. Although hyenas' nearest relatives are the cats and viverrids, of which mongoose are well-known members, they resemble dogs in their hunting and fighting methods. However, their family life is most like that of cats, and in their habit of defecating together at a communal latrine they resemble some mongoose.

While hyenas may be perplexing in the evolutionary origin of some of their behavior, the structure of their genital organs is even more so. Females mimic males so closely that it is difficult to distinguish the sexes, and in ancient times it was thought that hyenas were hermaphroditic. Hyenas display their organs to each other, particularly when greeting after a separation. Selection obviously favored some prominent structure for both sexes in this context and the logical choice was the male reproductive organ.

Long thought to be mere scavengers and killers of defenseless young, hyenas are in fact Dr. Jekyll and Mr. Hyde characters. During the day they are for the most part rather timid garbage collectors, a task

in which they are frequently aided by vultures. Like many predators, hyenas are consummate vulture watchers. Keeping the park under aerial surveillance, vultures are often the first to discover a carcass, and the sight and sound of them plummeting down, the air rushing through their pinions, alert predators to a possible meal. The white-backed vultures and Rüppell's griffons, with bare necks, take on the bloody job of stripping the meat; the lappet-faced vultures clean the bones, tearing at them with their broad, powerful beaks; and the hooded vultures peck like chickens at bits of dried blood and scraps around the carcass. Into this hissing and squabbling avian melee, a hyena sometimes dives and with furious snaps chases the vultures off. Armed with bone-crushing conical teeth and with such potent digestive juices that they can extract organic matter from bone, hyenas convert a skeleton to nothing but white fecal droppings composed of calcium. At other times, too, hyenas patiently hover around lions on their kill, waiting to drag the remains away, a sight which more than any other has created their public image.

But, as Hans Kruuk first described in detail, hyenas shed their air of subjugation at dusk, when they metamorphose into bold and powerful hunters with a liking for wildebeest and zebra, animals large enough to satiate at least a dozen members of their hungry clan. They chase their quarry over long distances in the manner of hunting dogs, often selecting a sick individual or a young one, biting at its flanks until the animal halts. They kill about two-thirds of their food themselves and scavenge the rest. One morning I came upon many hyenas surrounding an old wildebeest bull, an unusual occurrence so late in the day. He stood there mutely, glancing vacantly into the distance as if indifferent to the hyenas disemboweling him. I think he was in shock; at least I hope so. There is an aura of inevitability in these killings, one realizes that they must occur, yet one does not develop an indifference to them. After they have run down their quarry—and only about one-third of their chases are successful—the hyenas bolt their food. They are emotional animals, given to maniacal whoops, grunts, and wild laughter that can be heard from at least two miles away.

Any lion in the vicinity is certain to be attracted by the racket, and consequently the hyenas often lose their meal. Though the lions win some rounds, hyenas frequently even the score. In fact, on the plains, if lions are on a kill at dusk the chances are equal that hyenas will appropriate the remains at some time during the night. For instance, one afternoon I found four gorged male lions by an eland. At twilight, during that brief period when the night sounds take over from those of the day, hyenas began to gather. Appearing abruptly one at a time, like dark creatures from the nether regions, they gathered until there were twenty-five of them. At 8:30 p.m. their circle drew closer and they whooped loudly, their bushy tails raised, but the lions ignored them. Five hours later, at 1:30 a.m., they tried again, roaring ominously only thirty feet from the kill. Grumbling to themselves, three lions left, but one remained. The hyenas waited. A sweep of my flashlight revealed a circle of burning eyes focused on the lion. Then, as if on signal, they repeated their din, and moved ever closer until the lion lumbered off, the loser in a game of psychological warfare.

Almost all that is known about the social life of hyenas was discovered by Hans Kruuk, a colleague with whom I shared many pleasant months in the Serengeti. Hyenas live in clans numbering as many as sixty or more individuals. It is a matriarchal society in which females are the leaders and dominant members; the females are actually larger than the males, an unusual condition among mammals. Hyenas provide a fascinating example of how the structure of a society can be influenced by local circumstances. In Ngorongoro Crater, where food is abundant all year, hyena clans occupy discrete territories in which they hunt and where their communal den is located. Like man, hyenas have more than one system of morality: one for dealings within the group and the other for strangers. Members of a clan share kills but outsiders are usually repulsed in pitched battles that may result in the death of an individual. A sedentary existence is not possible for many of the three thousand hyenas that live in the Serengeti. The erratic wanderings of the migratory prey create periods of famine

on the plains where hyenas prefer to live. But hyenas have solved the problem in two ways. Some commute. Leaving their young behind in the den, the adults travel twenty, thirty miles until they find wildebeest or zebra. Then, perhaps three to four days later, they return to home base. Hyena mothers do not suckle the young of other litters, even though all may occupy the same den, and at certain seasons it is common to see lean youngsters haunting the den entrance waiting for the return of their mothers' foraging expedition. Some clans break up in times of prey scarcity and members follow the migrating herds for several months. Temporarily freed from having to assert territorial claims, such hyenas later meet strangers amicably and hunt with them, a good indication that possession of a piece of turf may have a remarkable effect on character.

CHEETAH

An area does not reveal itself all at once, and not until one has moved through it many times, viewed it in all its moods, by night as well as by day, does one begin to see it with a heightened consciousness. At first the Serengeti seems simple, an area composed of woodlands and plains, but with familiarity one notices the more intimate aspects and becomes acquainted with the other creatures that share this wilderness. Of course, driving around insulated in a car, one's senses dulled by the noise and stench of the engine, is a poor way to feel the land. Yet it is necessary, for the animals have learned to accept a vehicle whereas to a man on foot they respond with flight. Still, I derived a quiet joy from meeting acquaintances on my daily rounds: Corky, the three-legged crocodile basking by his usual algae-covered pool; George, the bull giraffe looking haughtily down on the car as I drive through his preferred corner of acacia woodland. And in the sparse shade of a whistling thorn, a cheetah female.

I knew that cheetah well but had never given her a name: her look of amber arrogance seemed to forbid such intimacy. When I first became acquainted with her in July 1967, she had a litter of cubs hidden in a kopje. A month later she brought them out of hiding. From then on they would have no fixed abode as they followed their mother on her daily hunts. She had three cubs then but one disappeared. The other two—both females—survived, wistful-looking creatures with a gray mantle of silky hair. For years now I have tried to fathom the purpose of this unique coat, which makes the young conspicuous and hence vulnerable, but I remain perplexed. Still, such trivial mysteries are piquant to ponder.

During the next few months I saw this family often, always just the three of them, for cheetah females avoid contact with their kind, even though several may use the same area without claiming a plot of land as their own. Males, too, generally travel alone, although two or three of them may become companions. The small groups of cheetah that one encounters are usually grown litter mates who have not yet split up. The female cared well for her cubs. When they lagged, she called them with a bird-like chirp followed by a soft chirr, surprisingly ephemeral notes from so large a cat. After a meal she licked their bloody faces clean, all of them purring loudly with contentment. Yet somehow their contacts seemed constrained; they lacked the intense and uninhibited desire to touch one another that is found in lions.

Their life turned upon a search for food. To find out how often cheetah kill, Kay or friends or I remained with the family during the daylight hours for twenty-six days late in the dry season. It was hot. Dust hung motionless over the somber plains, and the golden bodies of the Thomson's gazelle shimmered. For hours one waited, while the cheetah rested or surveyed the horizon. But one stayed, seduced by the expectation of that one brief moment each day when indolence explodes into action, when the stored energy is expanded in one glorious burst of violence. One day as I watched, the female suddenly became aware that several gazelle had drifted to within seven hundred feet of her. She tensed and rose slowly, the slight awkwardness of her repose transformed to elegant motion. With their trim waists and long slender legs, their deep chests and small heads, cheetah are the most atypical of the cats, built for speed rather than power, greyhounds with the coat of a leopard. Walking with head lowered, she approached her prey, partly screened by swaths of grass. Her cubs remained behind, eager spectators watching her with craned necks. She halted. One gazelle had its head raised and she waited tensely until it grazed once more, now only four hundred feet away. Another advance and suddenly she sprinted, her supple back coiling and uncoiling like a spring, faster, faster, her legs a blur, intent on a female gazelle that was somewhat separated from the herd. The ga-

zelle bolted in a flat gallop, but the cheetah gained, running at perhaps sixty miles per hour, puffs of dust rising where her paws fleetingly touched. With death imminent, the gazelle began to zigzag, but the cheetah followed each tortuous turn and suddenly flicked out a forepaw. Hit on the thigh, the gazelle crashed on its side, the cheetah lunged in, grabbed its throat and four minutes later death came through strangulation. Though panting from the exertion of the chase, the female dragged her quarry to a shady spot, followed closely by her eager cubs. She chewed through the hide, exposing the meat, then let the cubs feed while she regained her breath.

During these twenty-six days the female killed twenty-four gazelle and one hare. However, not all chases were successful. When cheetah pursue small fawns they almost invariably catch them, but their success drops to about 50 per cent when adults are the intended victims. When a cheetah is moving at high speed, even a slight miscalculation while a gazelle zigzags causes the cat to lose ground, and lacking stamina for a long chase, it must give up. Not surprisingly, cheetah select fawns whenever possible. The selection process is highly complex, and the whole success of the hunt depends on a series of rapid decisions that must be based on the vulnerability of the prey. Any cheetah which fails to adhere to certain rules will surely suffer. Size of prey is one obvious criterion. Being light of build and having relatively small and weak jaws, a cheetah cannot readily subdue prey weighing much more than itself; that is, animals of some 110 to 130 pounds. Impala, gazelle, reedbuck, and others of that size are, therefore, the preferred prey. To avoid wasting energy, a cheetah must select those animals which, according to its experience, are easiest to catch —the sick and the young; it must choose those which can be approached most closely—ones whose visibility is hampered by being behind brush, by having their head lowered while foraging, and so forth; and it must make certain that the potential quarry is unlikely to elude capture by disappearing into the anonymity of a herd. Cheetah not uncommonly abort a stalk because they are unable to make a selection.

Even after a kill has been made the cheetah is not assured of a meal. Lions and hyenas sometimes hear the dying bleat of the prey or spot vultures descending to the carcass. For example, a female once lost three of her twenty-five kills to these predators. Bounding up, they appropriate the kill while the timid cheetah retreats, moaning angrily to herself. I have even seen a phalanx of vultures advance ominously on foot, drawing ever closer until the cheetah vacated its kill. Because of such harassment, cheetah are nervous eaters, alternately snatching a few bites and glancing around until finally they abandon the remains with no attempt to save them for another meal.

In early December, after spending five months within an area of about twenty-three square miles, the cheetah family vanished. Its disappearance coincided with that of the Thomson's gazelle, which swarmed back to the eastern plains at the onset of the rains. As I also found in subsequent years, most cheetah migrate with the gazelle and for months not a single one visits Seronera. I searched for the family on the plains but without avail. Then on June 26, 1968, it reappeared in its old haunts. Kay and I were delighted to see them again. A year old now, the cubs were almost as tall as their mother though they were lankier and had a small ruff on the nape. Cubs must learn how to hunt before becoming independent, and I was curious to see what progress they had made. Not much. One day I watched a cub pursue a gazelle fawn. With a swat of its paw it bowled the animal over, again and again, without managing to grab it. Finally the female ran up and killed it. She then took the initiative in three subsequent hunts.

On October 17, when the cubs were just over fifteen months old, Kay saw the family in the evening, together as usual, resting contentedly. But by the next afternoon the two cubs had permanently separated from their mother. This dramatic and sudden break, this abrupt transition from dependence to full independence, surprised me, for it was unheralded by any form of behavior that I could in retrospect detect. Social ties among lions and leopards are severed gradually, preceded by tentative solitary excursions on the part of the young. Never again to my knowledge did the two cubs visit their mother even though all hunted in the same area and occasionally saw one another in the distance. At such times they haughtily ignored one another as if they were strangers. Unable to hunt efficiently, the two sisters grew lean and their pelvic bones protruded. But they survived and in December disappeared again. 

One day in February 1969, as I hiked across the plains near the Barafu Kopjes some twenty-five miles east of Seronera, I met a cheetah. She was lying on her side, head raised, glancing past me in an imperiously detached manner as if I did not exist. It was one of the sisters, now separated from her sibling. I sat down a hundred feet from her, rather pleased to have found a companion on this immense plain. Suddenly she jerked to attention and then I too heard the bleat of a gazelle fawn. It appeared over a rise, together with four adult gazelle, closely pursued by two jackals. Sprinting past me, the cheetah knocked the fawn down and quickly grabbed it: she had learned to hunt efficiently. While she ate, I moved toward her inch by inch on my elbows and knees until I was but fifteen feet away. She glanced at me guilelessly but did not flee. We stayed side by side for over an hour, once only ten feet apart, before she walked off, leaving me alone to exult over the perfection of our meeting.

By late April 1969, much earlier than usual, this female was back at Seronera. Her sister arrived in May and her mother returned in early June, accompanied by three new cubs, but one of them was ill and it soon vanished. In June, my interest focused on my friend from the plains, for she was obviously pregnant. On about July 10, at the age of two years, she had her litter. I found it two days later in a patch of tall grass, four tiny blackish cubs, eyes still closed and weighing less than a pound each. Their mother was just transporting them to a thicket several hundred feet away. She carried them there one at a time, and, after the last one had been moved, she returned twice more and intently sniffed around the lair as if checking to see that none had been forgotten. I checked their progress only occasionally, afraid that constant disturbance would cause her to abandon them. By watching the litter of her tame but free-living cheetah, Joy Adamson found that cubs can stand up by the age of nine days, that their eyes may not open until eleven days, and that they can walk at twenty-one days. The cubs remained hidden until, on August 19, at the age of almost six weeks, the mother led them to their first kill. The gazelle was not quite dead and the cubs were frightened of it, jumping back each time it kicked, according to Simon Trevor, who witnessed the event. Another generation of predators had begun its hunting lessons.

One cub of this litter disappeared in mid-August, another in September, and soon after that my study came to an end. Though I had learned much about cheetah, I still felt that I did not know them. The Serengeti area, ten thousand square miles in size, contains perhaps no more than 200 to 250 cheetah, and such low densities are typical of most other reserves too. Though cheetah lose about 50 per cent of their cubs, they are quite prolific and many are raised. They have reached a peak of evolutionary efficiency in their hunting. Then why are they so sparse, why is the species balanced so delicately between security and extinction? Their refusal to breed in captivity puzzles me too Thousands of cheetah have been kept in zoos and by Eastern potentates who trained them to hunt antelope, yet only about a dozen births have been recorded. Looking into their haunting eyes, I am aware that the cheetah remains an enigma.

# LEOPARD

The leopard is a solitary and secretive inhabitant of thickets, an animal of darkness. These characteristics have been both its salvation and its doom. By remaining out of sight, it has been able to survive near dense human habitation, areas from which other large cats have long vanished. But, whereas the lion invokes good will by displaying its indolent, seemingly carefree nature, the leopard conveys visions of a nocturnal marauder, of a cold, detached personality. Friendliness leads to friendliness. By virtue of its being so withdrawn the leopard has received little sympathy from man and now desperately needs it. Persecuted for its lovely hide, it maintains only a precarious foothold where it was formerly abundant. Every woman who needs to satisfy her vanity with a leopard-skin coat should first contemplate the exquisite beauty of this cat in repose. "The leopard's coat is like the dapple of firelight and dark on the forest floor, his eyes are the pale gold of the hunter's moon," in the words of Evelyn Ames. Leopard coats surely look best on leopards.

The African leopard has never been studied in detail and the notebooks of biologists contain only hints of its private life. Some seven leopards usually roamed the seventy-five square miles around Seronera. A male and two females remained for the three years of my study, one female disappeared, another was mauled to death by lions, a new female settled there, cubs grew up and left, but in spite of these changes the size of the population stayed about the same. I knew Scab-ear best. Not a graceful name for such a graceful creature, yet it described one of her distinctive features: flies attacked her ears so persistently that they were almost hairless and frequently swollen and encrusted with blood. She roamed along the watercourses that wound into the plains, a long and narrow range, some nine miles between the farthest points. Like others of her kind she was solitary. I never saw her meet the other females with whom she shared her range nor the male that entered the western corner of it.

Through frequent contact with vehicles and airplanes, Scab-ear had become inured to them. One day she killed two bat-eared foxes beside a bus filled with visitors; another day she pounced on the moving tow cable of a glider and was lifted ten feet into the air before she released her grip. But usually I found her sprawled on a branch, her luxurious tail dangling down and swaying gently in the breeze. Quite often a gazelle or reedbuck was wedged securely among the branches too, out of reach of prowling hyenas and lions. With a kill in the larder, a leopard can eat at leisure, for as long as three or four days, until all the viscera, meat, and skin have been devoured.

Leopards no doubt have the most varied diet of any large African cat. Records from the Serengeti include hyrax, baboon, guinea fowl, and python. I found four European storks stored in trees, a sign that these winter visitors are rather naïve about predators. Zoologist Richard Estes knew of a leopard in Ngorongoro Crater with a predilection for jackals, killing at least eleven of them within a period of three weeks. But their preferred prey are animals the size of impala, duiker, and wildebeest calves. Thomson's gazelle were favored items around Seronera and I soon noted that male gazelle outnumbered females among the kills by a ratio of at least two to one. When gazelle venture to a waterhole and suddenly sense danger, the females and young flee immediately whereas males often tarry a moment—a fatal hesitation. Males are also more careless than females, being more inclined to leave the safety of the short-grass areas to wander along thickets, through which leopards hunt. Thus leopards do not select males; rather males select against themselves by becoming vulnerable.

The leopard is considered a consummate hunter and indeed it glides toward prey with such smoothness that one is barely conscious of movement, a perfect combination of "beauty and utility, artistic and technical perfection," as Konrad Lorenz phrased it. But success is measured by kills not aesthetics, and on this basis the leopard is less adept than the cheetah. Of the nine stalks I observed only one culminated in a successful rush. During the rainy season, when prey was scarce, one cub out of a litter of two starved to death, for its mother was unable to capture enough for all.

I saw Scab-ear infrequently and my notes on her read like the intermittent family news of a distant friend. She was alone in the autumn of 1966 but sometime in January 1967, she had a litter. How many cubs she had I do not know, but one survived and it often peered from the hollow trunk of a sausage tree in which it spent the day during the early months of its life. The youngster, a female, grew rapidly and at the age of thirteen months she began to roam by herself. The physical ties between Scab-ear and her offspring loosened gradually but the emotional ones persisted for a time. Meeting after a separation, mother and daughter went through an ardent reunion, rubbing cheeks and licking each other, and it was at moments such as this that I realized the leopard's cold exterior hid a warm temperament. I never learned just when the young female began to capture prey on her own for she often shared a kill with her mother, but in August 1968, I found them three hundred feet apart, each with a gazelle in a tree. Not long after that the two ceased to associate. At the age of about twenty-two months the cub was on its own. And three months later Scab-ear conceived again. She had her litter in the dark recesses of a kopje in late May or early June 1969, two cubs this time. As I am writing this, nearly two years have passed since I last saw these cubs. They should now be independent and I wonder if they still pad around Seronera in the dappled shade of the fever trees or if they have sought wider horizons.

LI

# N

The first impression of an animal is often the one that remains most clearly in the mind. The day after I arrived at Seronera I found two lionesses at the edge of the plains, their tawny pelage so much a part of the golden *Themeda* grass in which they sat that to me any yellow sward is now somehow incomplete without a lion. Their eyes were fixed on a herd of Thomson's gazelle, their whole bodies straining toward it, as if willing it to approach. Moving in single file, the herd drew closer and the lionesses parted, one to the right, the other to the left, mere wisps of wind among the tall stalks as they snaked into position. Oblivious of danger, the gazelle continued along a well-worn trail until they were between the lionesses. A breeze stirred the midday lull and suddenly the gazelle scented lion. Leaping and twisting they scattered and raced away, still ignorant as to where the enemy lurked. One gazelle rushed toward a hidden lioness, and spotting the motionless form too late to change course it leaped high above the grass in a desperate attempt to escape. But with exquisite timing the lioness reached up and with gleaming agate claws plucked the animal out of the air. Incidents such as this help deflate one's sense of superiority, the feeling that man is somehow graced in all respects with greater power than other creatures. Every species has those attributes which best suit it for a certain type of

existence, and looking at my own body so lacking in weapons, other than cunning, I can only marvel that man survived at all.

A study is most interesting and satisfying when one deals with known individuals. As lions lack a distinctive coat pattern, such as the stripes of zebra and splotches of giraffe, I realized that there would be difficulties in recognizing many animals with certainty in the vastness of the park. I solved this problem in two ways. Selecting a limited area around Seronera, I concentrated on identifying every lion in it by such characters as torn and notched ears, scars, and other blemishes. Most lions lived permanently there and soon became acquaintances whom I knew at a glance. There was, for example, The Old One, whose worn canines, drawn look, and heavy tread showed that she was the local matriarch. One-ear and One-eye, both lionesses, often traveled together as if their deformities were a basis for their bond. And there was Flop-ear, the quintessential lioness, a big, bold, beautiful animal in the prime of life. Also using the area were three males with huge manes, an aristocratic trio who radiated an almost palpable power. Like the Three Musketeers they roamed together through their domain, but to name these sedate, shaggy beasts Athos, Porthos, and Aramis was not befitting and I called them prosaically Black Mane, Brown Mane, and Limp.

To help plot the lions' movements, I placed ear tags on 156 of them. Driving close to a cat, I fired a syringe from a carbon dioxide-powered gun into its thigh. The muscle relaxant succinylcholine chloride is injected on impact and about five minutes later the lion rolls onto its side. First I clamped a colored and numbered metal tag into each ear, then sometimes took a blood sample for later disease studies, and finally withdrew into the car, watching the lion's recovery carefully. Usually it raised its head, fully alert, within twenty-five minutes, quite oblivious of the tags. In most instances the whole operation went so smoothly that other pride members either ignored me or watched the activity from as close as thirty feet. However, once a lioness died, to me such a traumatic experience that I nearly gave up this aspect of my work. Except for the sharp, sudden pain with the impact of the syringe, the lions suffered little discomfort. Yet I came to detest tagging and after the first few necessary months did little of it. In the beginning it did give me atavistic pleasure to squat by a lion and slide my palm over its sleek, slightly oily hide, to run my fingers through its tangled mane, but with time I developed such a feeling of tenderness for these beasts that I found it abhorrent to disturb them in any way.

Tagged lions appeared occasionally near Seronera, where visitors saw them and objected to the adornments. Nobody seems repelled by large numerals painted on the hides of elephants or garish collars placed around the necks of wildebeest, but a tagged lion is a desecration. As Evelyn Ames noted: "Lions are not animals alone: they are symbols and totems and legend; they have impressed themselves so deeply on the human mind, if not its blood, it is as though the psyche were emblazoned with their crest." I had empathy with the protesters—tourists, park wardens, hunters—but I also needed the information that tagged lions would provide.

For example, without tagging, I would not have known male No. 57. I first met him in November 1966, at Musabi, a small plain in the western part of the park. He was about two and a half to three years old, a rather scruffy fellow with tufts of mane sticking untidily from his cheeks and a rakish ruff down his nape. Like almost all males of his age he had abandoned his pride, maybe forced out by the adults or simply overcome by wanderlust, and he now roamed alone, a nomad with neither home nor companions. I saw him twice more around Musabi, but in March 1967, after the migratory herds had moved to the plains, he appeared there, fifty-five miles east of where I had tagged him. He was still on the plains in June, sometimes alone, at other times with lions he had met casually. Though not antisocial, he had a restless spirit and always left his new-found acquaintances. However, June 12 was a momen-

tous day in his life. He had rested all day on a rise and at dusk set out purposefully on some mysterious errand. At 9:25 p.m. he met a male of his own age and they walked on together, and by some intangible process they cemented a friendship which lasted until death. They spent July to about December in the woodlands near Musabi and in January 1968 returned to the plains. I last saw them there in late May, now powerful, handsome males with full manes. Where they spent the dry season I have no idea. But I do know that on November 9, 1968, they were northeast of the park, outside of its boundaries. I know that a Land Rover or maybe a Toyota drove close to them, stopped, and a man with a rifle stepped out—assuming he obeyed the law. The males had no fear of cars and even a person on foot caused only uneasiness. Perhaps they stood watching, perhaps they lumbered off to continue their rest elsewhere. In any event, a bullet slammed into male No. 57. The outfitter of the hunt sent the ear tag to me. It was a generous gesture, yet I cupped the blood-encrusted silver tag in my palm with a terrible sadness. I would rather have retained my vision of male No. 57 wandering through his kingdom, the grassy plains, the hollow vastness of the sky. Now I see him nailed to a wall, staring glassy-eyed, his teeth bared in supplication.

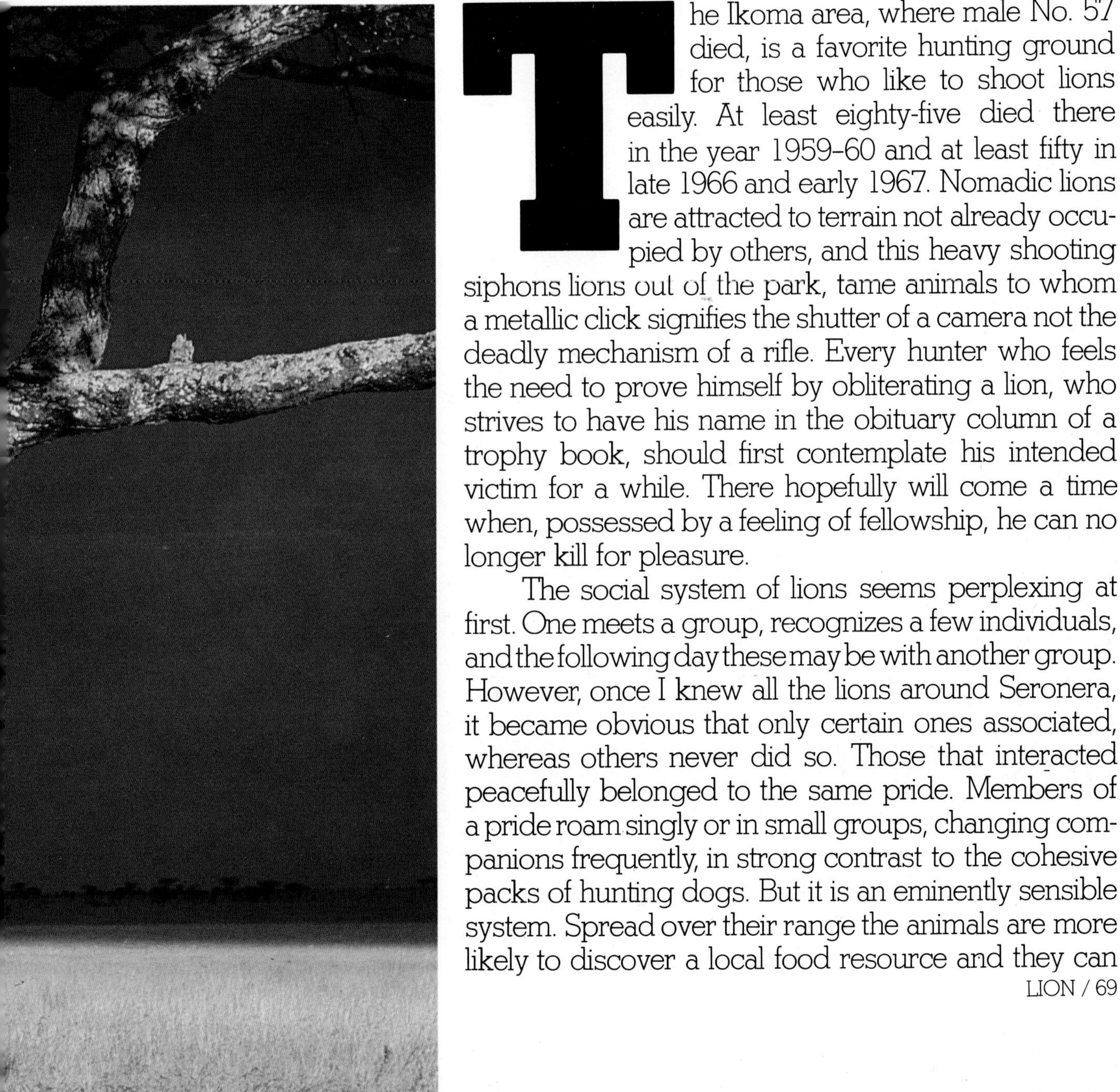

The Ikoma area, where male No. 57 died, is a favorite hunting ground for those who like to shoot lions easily. At least eighty-five died there in the year 1959–60 and at least fifty in late 1966 and early 1967. Nomadic lions are attracted to terrain not already occupied by others, and this heavy shooting siphons lions out of the park, tame animals to whom a metallic click signifies the shutter of a camera not the deadly mechanism of a rifle. Every hunter who feels the need to prove himself by obliterating a lion, who strives to have his name in the obituary column of a trophy book, should first contemplate his intended victim for a while. There hopefully will come a time when, possessed by a feeling of fellowship, he can no longer kill for pleasure.

The social system of lions seems perplexing at first. One meets a group, recognizes a few individuals, and the following day these may be with another group. However, once I knew all the lions around Seronera, it became obvious that only certain ones associated, whereas others never did so. Those that interacted peacefully belonged to the same pride. Members of a pride roam singly or in small groups, changing companions frequently, in strong contrast to the cohesive packs of hunting dogs. But it is an eminently sensible system. Spread over their range the animals are more likely to discover a local food resource and they can adjust group size to prey size: when gazelle, which provide lions with little more than an *hors d'oeuvre,* are the principal quarry, hunting groups are smaller than when wildebeest are sought. In general, a pride consists of one to four males, up to about fifteen lionesses and cubs of various ages. For example, late in 1966 the Masai pride near Seronera consisted of three adult males, six adult females, a subadult female about three years old and ten cubs all less than one year old, a total of twenty-two. Another pride contained thirty-seven at one point, a third pride merely four.

To understand lion society fully it is necessary to consider females and males separately. The lionesses stay together for life, generation succeeding generation. A few young lionesses leave the pride when they are two and a half to three and a half years old to lead a nomadic existence. Such emigration, coupled with occasional deaths, tends to keep pride size stable. But other young lions maintain their family ties, which are so strong that no stranger is accepted. Watching lionesses, I sometimes wondered for how many years they had inhabited their ancestral ground, for how many generations mothers had passed on to daughters special pride knowledge, such as the boundaries of their range, the best ambush places, the most secure thickets in which to have cubs.

Males are only temporary pride members, staying a few months to a few years before they move on. Of twelve prides only three still had the same males at the end of the study as at the beginning. C. A. W. Guggisberg told me that in Nairobi National Park males were able to retain a pride for six years on two occasions. Sooner or later they leave, some abandoning the pride of their own volition but others being forcefully evicted by nomadic males. Almost all male cubs in a pride become nomadic by the age of three and a half years and they then roam, ready to expel pride males if these show signs of weakness. However, Stephen Makacha observed that in Manyara National Park a young male joined forces with his father to rule the pride in which he was born. Nomads often travel in small groups. Such animals are frequently brothers or pride mates that grew up together, but some become companions later in life, as was the case with male No. 57 and his friend.

Male No. 57 died before he could rule a pride, but another group successfully made the transition from a nomadic to a settled existence. I first met them in March 1967, at the Moru Kopjes south of Lake Magadi. There were five males, about three and three-quarters years old, sturdy animals with the cocky expression of young manhood, or rather lionhood. During the rains that year they wandered widely over the eastern plains, retreated to the Moru Kopjes for the dry season, and in December were once more on the plains. But this time they homesteaded a piece of land around Naabi Hill. Since the plains are almost devoid of prey for part of the year, prides cannot establish themselves permanently, leaving hundreds of square miles unclaimed. The males were there until May, patrolling the rises and swales, their roars challenging the night: "This land is mine, mine, mine." But the dry season forced them to retreat. On September 10, 1968, several members of the Lake Magadi pride harassed a solitary bull buffalo. Silver Mane was there and so were Scar Nose and several females and cubs. These two males, as well as a third one, had been with the pride for at least two years. Suddenly the five nomads arrived at a trot, roaring hoarsely, the embodiment of a strength which nothing could resist. Pursued, the pride fled through a thorn thicket and toward a ravine where I could not follow. From that day on, the Lake Magadi pride lionesses had new overlords. And their former ones? Two I never saw again, but in January 1969 I met Scar Nose on the plains, alone, now a harried-looking nomad who along with his territory had also lost his quiet dignity.

The life of a male is insecure. Fewer than 10 per cent reach old age. Three-fourths die violently, caught in poachers' snares, shot by hunters, and killed in fights with others of their kind. Perhaps this is just as well. Males are sometimes sick or wounded and slowly they starve, the final resistance of their emaciated bodies

concentrated in their solemn eyes. Pride membership is a form of life insurance in that each animal can sustain itself at the kills made by others. With few exceptions, old males lack such membership, and alone, shrouded in their regal past, they sink into oblivion, their future in the hands of hyenas and vultures. An animal probably knows no despair, yet these males filled me with unutterable sorrow, the sorrow of our human soul when faced with the indifference of nature.

Each pride confines itself to a definite territory in which strangers are usually not welcome. Fifteen square miles may be enough space for a pride's needs when prey is abundant, but some use as much as 150 square miles. Although territories overlap extensively, neighboring prides seldom meet and nomads usually manage to avoid contact with the owners too. The chance of confrontation is reduced in several ways. Prides do not use all parts of their territory equally but spend much of the year in what can be called a focus of activity, an area with ample food and water that is left mainly in times of shortage. These foci overlap little or not at all. A pride may not even be aware of intrusions into the peripheral parts of its territory and nomads sometimes reside there for weeks as squatters.

Prides also advertise their presence. Males patrol their domain, and as one of them ambles along he halts occasionally at a tuft of grass or shrub and first languorously rubs his face in it, eyes closed as if in ecstasy, before swiveling around and with raised tail squirting the site with a mixture of scent and urine. This pungent calling card conveys to a stranger not only that the area is occupied but perhaps how recently the owner passed. A more emphatic signal is the roar. As Laurens van der Post noted, "it is to silence what the shooting star is to the dark of the night." At first a moan or two fills the void, then several full-throated roars that die away in a series of hoarse grunts, leaving the earth shaken and subdued. When a pride roars communally it surely is one of the most impressive choruses in nature. The meaning of a roar depends on who hears it. To a pride member, secure in the knowledge that it belongs in the area, a roar communicates the location of companions, to an outsider it marks a place to be avoided.

Tremendous strength, long canines, and sharp, hooked claws are all specializations that have made lions supremely well adapted to overpowering and killing prey—as well as one another. This makes encounters dangerous, especially since lions have few inhibitions about using their weapons at close quarters. However, battles are avoided whenever possible. A male may instead try to impress and intimidate a stranger by displaying his fine physique and voluminous mane while strutting in front of him. The sex of the intruder also influences the vigor of attack. A male vents his antagonism mainly on other males, whereas he may tolerate a lioness, especially if she is in heat; a lioness in turn accepts new males, a sensible gesture, for without it replacement of pride males would not be possible. Intruders are chased with much roaring, but it always amused me that the pursuers usually took care not to run fast enough to catch anyone.

Yet no system is foolproof and the histories of the neighboring Masai and Seronera prides illustrate what may happen when the balance between aggression and restraint is upset. During my first year no dramatic events interrupted the level tenor of their life, even births and deaths being only a part of the inevitability of things: The Young Female had her first litter at the age of almost four years, and Female Q, as I designated her, permitted her single cub to starve. Alan Root, who was filming my project, watched the two prides meet once and later so did I, but the encounters were brief. On September 1, 1967, the males of the two prides were within a mile of each other in an overlapping part of their range. The following morning, at 6:35, I found one of the Seronera males encrusted with blood, his body covered with punctures and cuts. A deep gash angled across his brow, closing one eye, and a fist-sized hole penetrated his chest. Tatters of his golden mane littered the area, mute evidence of a titanic struggle. He breathed with difficulty. Suddenly Black Mane of the Masai pride emerged from a nearby thicket and slowly walked up to the vanquished one. He gazed down at him and received a faint growl and a baleful glare. Then Black Mane returned to the remains of a zebra, whose presence may have been one cause for the dispute. Almost imperceptibly life ebbed from the wounded male. And when at 8:35 his breathing became feeble, his bladder emptied, and finally his pupils grew very large, I felt almost ashamed for intruding on the freedom of his last moments as the amber fires faded from his eyes.

His death had dramatic effects on both prides. Looking back at the incident and its consequences has helped me reach a deeper appreciation of the complexity of lion society. For, with the death of that male, the males from three other prides penetrated into his former territory even though his companion was still there. Alone now and unable to keep intruders out, he lost his self-assurance and slunk around his domain; finally two males of the Kamareshe pride, whose range adjoined that of the Seronera pride to the west, drove him out. I never saw him again. The lionesses suffered too. An unknown assailant pursued three cubs and one at a time bit them to death, the bodies and tracks in the dusty earth revealing the grim tale. Another morning the Kamareshe males found three cubs hidden beneath a fallen tree and killed them. One male ate the viscera of a cub and the other carried one off with him like a trophy. I waited by the third cub for the return of its mother. At dusk she came. I was not certain what to expect, certainly not a wild expression of grief but perhaps some sentiment. Instead she ate the cub, and as I sat there in the dark listening to her crunch the bones of her offspring, I could only conclude that it is difficult for man to return in imagination to the simplicity of a lion's outlook.

The Seronera females now associated with various males but seemed unable to attract permanent masters. These males were careful not to meet. Feeding with the lionesses along a river one night, the Masai males suddenly jerked to attention, then fled roaring. Arriving from downstream were the two males of the Nyaraswiga pride. But, when they heard the roars, they too fled, with the result that both sets of trespassers bolted in opposite directions. Sensing that the other males were in retreat, the Masai males turned and chased them. The Nyaraswiga males sneaked silently across the river and the pursuers went roaring downstream for another mile. 

The various males spent the rainy season with their respective prides, but the following dry season they continued to indulge their dilettantism. The Masai males spent progressively more time away from home and finally, in August 1969, almost two years after the death of the Seronera male, they left their pride and took over the Seronera lionesses. Now the Masai lionesses were unattended, ignominiously deserted. But not for long. Almost three years earlier I had met two nomadic males on the plains, No. 58, who was a rather ragged middle-aged fellow, and an old male with a long, haggard face and weak hindquarters yet with a magnificent mane. Also on the plains were two solitary lionesses. These four lions as well as another female met early in 1967 and settled down in a shallow valley on the plains. They spent the dry season in the woodlands, then they trekked back to their valley, repeating this cycle annually like many other nomads. The summer of 1969 found them at the edge of the woodlands within the territory of the Masai pride. When the Masai males switched their allegiance to the Seronera lionesses, male No. 58 and his friend grasped the opportunity and abandoned their lionesses for the security of an established pride. Considering their advanced age, their tenure with the Masai lionesses may be brief, but they may have left their imprint. Just before I left the Serengeti they were courting with two of Flop-ear's offspring, whom I had watched as woolly toddlers, as exuberant youngsters, and now as adults. The study had come full circle.

Males are often pictured as indolent freeloaders who leave the lionesses to do the hunting and raise the cubs, as if their sole function is to procreate. The tribulations of the Seronera pride showed that males do indeed have an important role in lion society: they help maintain the integrity of the territory, thereby providing lionesses with a secure place in which to have cubs. The Masai pride was over twice as successful in raising its young as was the Seronera pride. Hunters who shoot male lions on the assumption that such animals are expendable might think in terms of the disruption and cub deaths their act may cause in the pride.

Males radiate "a kind of lazy, lordly power born of the carelessness of authority," as Maitland Edey phrased it. And indeed they give the impression of being despots who take what they want, whose existence lacks all subtlety. Yet, as I slowly learned about lion society, the males were the ones that provided me with some of the finest moments of revelation. Pride males are rather reticent about expending their energy in hunting. For instance, when a male fed on a gazelle the odds were 76 per cent that he had taken the meat from a lioness, 12 per cent that he had scavenged it from another predator, and only 12 per cent that he had caught it himself. Setting off at dusk on a hunt, the lionesses are in front, tensely scanning ahead, the cubs tag playfully behind, and the males bring up the rear, walking slowly, their massive heads nodding with each step as if they were bored with the whole matter. But slothfulness may have survival value. With lionesses busy hunting, the males function as inadvertent guards for the cubs, protecting them particularly from hyenas. Bulkier and adorned with a voluminous mane, males are also more conspicuous than lionesses, and in a business which for success places a premium on being hard to see and agile, it is just as well that males stay out of the way.

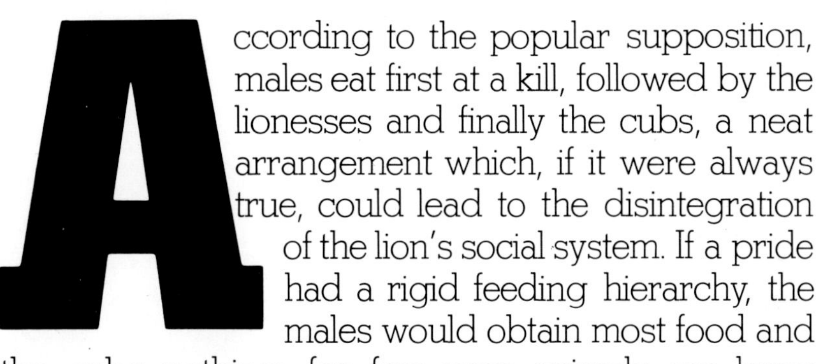

According to the popular supposition, males eat first at a kill, followed by the lionesses and finally the cubs, a neat arrangement which, if it were always true, could lead to the disintegration of the lion's social system. If a pride had a rigid feeding hierarchy, the males would obtain most food and the cubs nothing, for few prey animals are large enough to satiate many lions. Of course males, being so much larger and stronger than lionesses, usually win any dispute over spoils. But there is no hierarchy among pride members in the sense of one animal being dominant and others accepting subordinate positions. Each fights for its rights and vigorously defends its share of meat by snarling and cuffing, behavior which makes another animal reluctant to attack. It is a system of peace based upon the amount of damage each can do to the other, a system based on a balance of power. Consequently all crowd around a kill flank to flank, "rumbling at each other with the collective menace of a volcano," as Robert Ardrey described it. With snapping jaws and flashing claws, lions assert themselves, and it is not coincidental that 6 per cent of the animals I studied had only one eye. The vigor of the disputes often seemed excessive, especially if there was ample meat, and I frequently had to remind myself that such behavior was adaptive or it wouldn't have persisted in evolution.

That pride males have no hierarchy is most clearly evident not at a kill but during courtship. A lioness in heat is closely followed by a male and the pair usually becomes separated from the rest of the pride, not because they prefer privacy but because they are more intent on each other than concerned with such mundane activities as hunting. Quite often another male is near the courting pair, resting a few hundred feet away, seemingly bested by his rival. To see what would happen if the courting male was temporarily removed, I tranquilized one on two occasions. The extra male claimed the female immediately. When the first male revived after a few minutes, he made no attempt to evict the usurper. Possession obviously confers temporary dominance.

Whereas hunting dog society emphasizes the welfare of young, lions are self-indulgent, seldom permitting the needs of cubs to conflict with their own desires. Of seventy-nine cubs that were born to the Seronera and Masai prides during the study, 67 per cent died:

| | NUMBER |
|---|---|
| KILLED BY LION | 11 |
| KILLED BY LEOPARD | 1 |
| KILLED BY HYENA | 1 |
| STARVATION | 15 |
| UNKNOWN (NOT STARVATION) | 25 |

Many cubs simply disappeared, often whole litters of vigorous youngsters, and I think that in some instances they were simply abandoned. Perhaps leaving them to die in the obscurity of a thicket, their mother returned to the pride's conviviality.



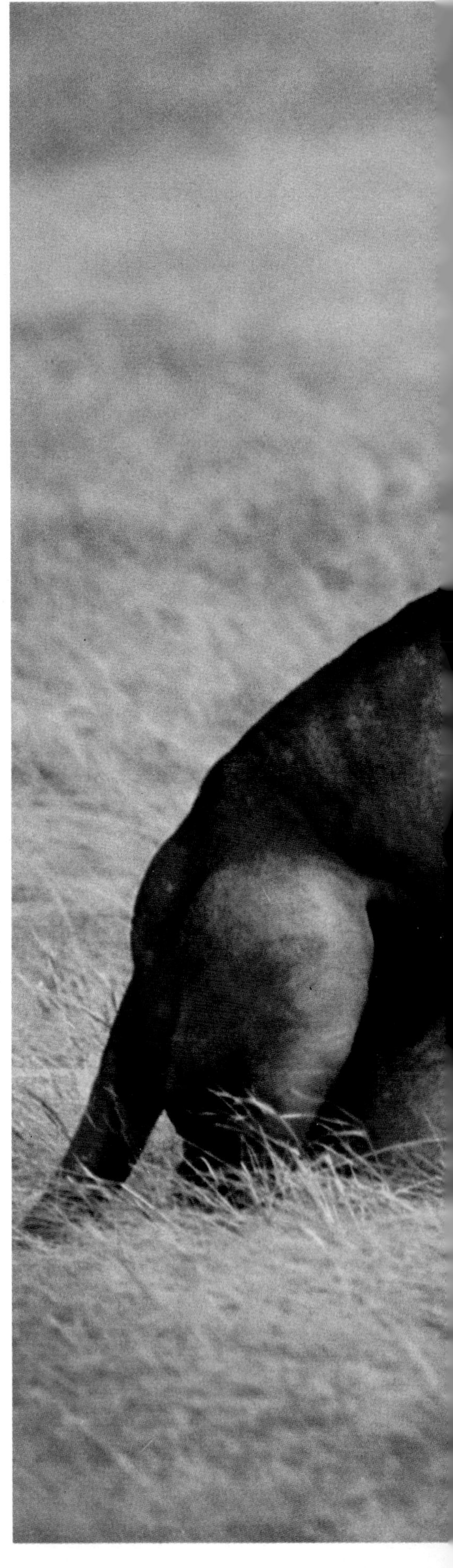

It seems rather paradoxical that cubs should starve in an area that contains such a huge wildlife concentration. However, cubs do not usually die because there is no food but because during the dry season they were unable to get a share of the small Thomson's gazelle, which are almost the only available prey around Seronera at that time. Adults tear a carcass apart and fiercely defend their portions, and it was depressing to see a starving cub totter to its mother, each rib sharply outlined beneath its unkempt hide, and receive a vicious cuff instead of a bite to eat. Very occasionally a mother may carry a gazelle to her cubs or hide the kill and lead them to it, but as often as not another lion steals the carcass.

A pleasant myth has it that a mother is helped in the care of her cubs by an "auntie," a childless lioness who hunts, guards, and does other chores. While it is true that a childless lioness may be around cubs, her motives are suspect and her presence is of little help. She may, of course, inadvertently provide some protection, but she does not lead cubs to kills or share food with them; rather, she deprives them of their share whenever she has the opportunity to do so.

Cubs always manage to obtain a meal from a large carcass if they are present when the kill is made. However, if they are far away waiting patiently in a thicket for their mother's return, they again have problems. Only after the lioness has gorged herself is she likely to fetch her cubs. By the time she walks perhaps a mile and then returns with her offspring there may be no meat left. And here is where a male inadvertently comes to the rescue. At first eating together with the lionesses, the male may suddenly drive them off the carcass and prevent them from taking the remains, often just the rib cage, neck, and head. But cubs are permitted to join him and in this way the latecomers are sometimes provided with a meal. Altruism in human terms has little place in nature. But I am filled with a sense of wonder when contemplating behavior which in my ignorance I would at first glance have dismissed as merely selfish and yet which is actually a subtle evolutionary product of benefit to the pride.

This annual period of starvation among the cubs presented me with a moral problem. Traditionally the park shot prey to feed such cubs. On scientific grounds I was against this practice for several reasons: the Serengeti is not a zoo, and, that being so, animals should be allowed to regulate themselves without interference whenever possible; the lion population was healthy, increasing in the park at a rate of at least 5 per cent per year, so did not need assistance from man; and the size of the Seronera lion population had stabilized around an optimum level, based on a subtle interplay between prey abundance and social factors. To increase the number of lions artificially would simply cause excess animals to emigrate and possibly be killed elsewhere. I also objected to shooting perfectly healthy prey, to disrupt their social system, to feed decrepit lions. Having few enemies, the lion must regulate its population in part by limiting the number of cubs that reach adulthood. The greatest good for the greatest number is one of nature's tenets and the individual must be sacrificed to that end. Yet these arguments caused me anguish, for when I saw a starving cub, its face drawn, its eyes vacant, my emotions were with the individual not with the race.

One cub I did rescue. Rameses' elderly mother was of the Seronera pride. Apparently unable to produce enough milk, she carried her dying cub to the remains of a wildebeest and abandoned him. Finding him there, an emaciated bundle of dank fur at the mercy of vultures waiting in the trees around him, I took him home, knowing that his life would have to be in a zoo, yet feeling that in a small way I was atoning for all those that were permitted to die. Gentle, friendly, exuberant, aloof, irascible, and fierce, Rameses fused these traits into a complex personality that enthralled us and enriched our understanding of his pride mates even in the three months we kept him.

When adults are lean and cubs starving one wonders why prides remain sedentary, clinging to a territory, while nomads are sleek and fat with the migrating herds on the plains. As the Seronera pride revealed, lions need a stable social environment in order to raise many cubs successfully. Such an environment is of advantage because males protect the area, secure denning sites are available, and lionesses feed, guard, and hunt for cubs communally. A migratory pride would lose some of these advantages, assuming that the members could remain in contact. Nomadic lionesses seldom raise cubs even though they often have them. And, in the final analysis, selection is for reproductive success.

---

Lionesses generally reach sexual maturity between the ages of 3 and 4 years. Of four young lionesses in the Seronera pride, two conceived for the first time at 3 years 6 months, one at 3 years 8 months, and one at 4 years 2 months. Males, too, mate for the first time between 3½ and 4 years, an age when they have a sudden spurt of growth in body and mane. A lioness in heat is most attractive to a male. He lounges beside her and moves whenever she does, walking very erect beside or just behind her. He may lick or nudge her gently when she stands, and, if she is receptive, she crouches in front of him after first rubbing herself sinuously against his body. As they copulate, she emits a rolling, ominous growl almost continuously. He then often grasps her nape with his teeth. Young cats become immobile when their mother holds them by the scruff, which makes it easier for her to transport them, and it seems likely that the neck bite of the male induces temporary passiveness in the lioness. He needs that extra moment of peace, for as soon as he advertises his climax with a prolonged harsh miaow, she whirls around with bared teeth and clouts him as often as not. A male mates repeatedly: one did so 157 times in 55 hours, once every 21 minutes on the average. Female cats are unusual in that repeated copulation is necessary to stimulate the release of eggs from the ovary. Since a lioness remains in estrus for three to five days and on occasion as long as two weeks, the male not surprisingly may relinquish her to another male after a day or two.

About 80 per cent of the courtships did not result in births. Sometimes a lioness was already pregnant or lactating when she mated, at other times she apparently failed to come fully into estrus, but in about a third of the cases she went through a typical cycle without conceiving. In addition, 17 per cent of the lionesses never seemed to come into heat. When several lionesses lacked cubs and one of them came into estrus, this stimulated others to do so too, with the result that several litters were born at about the same time. It is not unusual to see a dozen cubs of about the same age in a pride and the communal care they receive no doubt benefits them all.

Cubs may be born during any month but at Seronera there were two definite peaks, one in January and another in August-September. What caused these? Analyzing my data, I discovered two points of interest. One was that the simultaneous breeding by several lionesses in a pride may effect a peak. If, for instance, cub survival was poor the previous year, perhaps because of a food shortage, then many lionesses would be ready to conceive at the same time, the animals stimulating one another to come into heat together. The second point was that environmental conditions may affect sexual activity. When the migrating herds arrived on their unpredictable treks, providing a superabundance of food, lionesses promptly came into estrus. Thus the timing of birth peaks is somewhat accidental and will vary from area to area and year to year.

Around Seronera, litters varied from 1 to 5 cubs with an average of 2.3. If a whole litter dies, the lioness may immediately come into heat again, and with the gestation period being 100 to 119 days she should have another litter within 4 months. However, the actual interval was 9 months, because she either failed to come into heat or did not conceive as often as expected. If, on the other hand, a litter survives, at least 18 months—and usually 22 to 26 months—elapse before the next litter is born. Though lions have a high reproductive potential, they control their population in part by physiological means, a more effective method than the cultural ones, on which man is mainly dependent. Birth control together with cub mortality is a major method of population regulation among lions: the prides around Seronera raised only about a quarter of the expected number of cubs.

C. A. W. Guggisberg told me the interesting history of Blondie, a lioness in Nairobi National Park. She had two cubs in July 1955, but both died. Conceiving again immediately, she had four more cubs and raised these successfully. Between late 1957 and mid-1958 she had two litters but abandoned them both. The same fate befell a litter born in March 1959, and another one born that July disappeared. But after this dismal record she raised two cubs between 1960 and 1962. After that she had no more young, and in 1967 she died, possibly of old age, being at least seventeen years old at that time. Blondie had seven litters in twelve years, of which she raised two—only six cubs. A species which lives a long time and has a fairly low death rate cannot afford to raise an indiscriminate number of young.

Lionesses hide their newborn cubs in thickets and kopjes and these remain there quietly when their mother is away but greet her with piercing miaows when she returns. I sometimes crawled into a lair to count the cubs. While enjoying the chance to handle the soft, woolly creatures, I was distinctly uneasy, for the mother or one of her companions might return and take a dim view of the intrusion, even though it was made in the name of science. Weighing a mere three pounds, eyes still closed, a cub is quite helpless at birth. Their eyes usually open after three to fifteen days. They are gray-blue in color, which slowly changes to amber when they are two to three months old. By the age of three weeks they can walk, though not well enough to follow their mother far, and should she decide to move to another lair she picks them up by the nape and carries them one at a time. Other pride members may visit the cubs. However, at times cubs see no other lions until their mother leads them to the pride when they are some five to seven weeks old. To be confronted with the powerful males and to compete for meat for the first time must be traumatic experiences to a cub.

One morning I watched as a Seronera pride lioness introduced her two small cubs to six older ones resting together in the shade of a tree. Excited about the newcomers, the large cubs bounded up, but the others fled, unused to strangers much less to such exuberant ones. One hid in a thicket all day; the other sought refuge behind its mother. When the large cubs inched closer, craning their necks with curiosity, the small one snarled and slapped at them until finally a cough from the lioness sent them scurrying away. But, after an hour of proximity, the small cub became courageous and cautiously ventured to within a foot of another cub before dashing back to its mother and gaining reassurance by vigorously rubbing cheeks with her. Three hours later it hesitatingly touched noses with one of the large cubs and after another one and a half hours it suckled beside one, the task of integration well advanced.

A cub's life is not just one of hardship, of cuffs and snarls, or abandonment and starvation. Most days pass pleasantly in play, sleep, suckling, and merely reveling in the intimacy of the pride. However, lions in daytime give as erroneous an impression of their life as would a group of people observed only after dark. Lions are active mainly at night and only then do they reveal fully their power and beauty; they assume a new and implacable mien, of boldness and tension. To find out how cubs and adults behave at night, I often stayed with them at that time, usually when the moon was full, because the use of artificial light would be an intrusion which might well affect the behavior of both predator and prey. For me each month revolved around the moon, and watching it grow from the first radiant sliver to brilliant fullness I became aware again of the stars and planets, the cosmic rhythm of life which one tends to ignore when sleeping within confining walls. One day, at 11 a.m., I met all the Masai pride lions except Limp far out on the plains, where they were resting by a waterhole. Five four-month-old cubs belonging to Flop-ear and Mrs. Notch were with the pride then. Knowing that the moon would be almost full that night, I decided to stay with the lions for twenty-four hours. Here was their routine:

The animals were "poured out like honey in the sun," as Anne Morrow Lindbergh has aptly described them, some on their backs, others on their sides, legs and bodies touching as if contact enhanced the pleasure of their repose. Only Scruffy, a young male, remained at the periphery, because Flop-ear was antagonistic toward him while she had small cubs. Occasionally a lion changed to another rest site, collapsing there abruptly with an audible sigh, its legs suddenly turned to rubber, utterly exhausted, it seemed, by the effort. Both at 12:30 and at 2 p.m. several cubs became restless and wandered aimlessly around before wending their way to Mrs. Notch's milk supply. There, as usual, they squabbled, trying to elbow each other aside, until Mrs. Notch, annoyed by the needle-sharp claws that raked her breasts, drove them off. Ignored by the adults, several cubs played gently. One bit a grass stalk and shook its head as if subduing an opponent; two others batted at each other with their paws; and another caught the flicking black tassel at the end of a lioness's tail and gnawed on it, no doubt assuming that this delightful appendage was specifically designed for its pleasure until a snarl convinced it otherwise. Their play was subdued and soon ceased.

Cubs play mainly when they feel secure, when neither starved nor gorged, when neither hot nor cold. Exuberant play is most common around dawn and dusk. Wrestling, chasing, and stalking may then follow one another in bewildering succession for as long as an hour. Small cubs wrestle much, but by the time they are six months old stalking has become more important to them and finally as subadults they seldom indulge in the intimate contact that wrestling provides. A lioness usually tolerates play-

ing cubs, even when these jump up against her, drape themselves over her face, tug at her paws, and otherwise make a nuisance of themselves. In fact she may reciprocate, slapping a cub hard yet with restraint or covering it with her body until it manages to squirm free. The males, on the other hand, are usually surly, rebuffing playful cubs or at most placating them with a cursory pat.

Play is a fascinating form of behavior. Everyone recognizes it, but it is difficult to describe and its functions remain obscure. Most play patterns in animals developed for purposes other than play. Lions, for example, borrowed their patterns from many different adult contexts: they stalk, rush, chase, and pull each other down as they would prey or another lion, they slap as they would in combat, and they drag sticks as they would a kill. Play gestures are exaggerated and their sequence may be different, but they are not new. The main difference between play and the adult pattern lies in the fact that the former lacks the corresponding emotional state: play-fighting cubs do not snarl with exposed teeth.

Play must have functions or it would not have evolved in so many mammals, but to define these is difficult. Play expends excess energy, it helps coordinate complex movement patterns in a maturing cub, and it is a friendly activity which perhaps helps strengthen social bonds within the pride. Also a cub that has learned to stalk in play will presumably catch prey more successfully later. But why do cubs have to practice in a playful rather than serious manner? And why do lionesses play? Perhaps through play cubs also learn the rules of pride life. A playful approach, with exaggerated bounds followed by a nip, is less likely to be misunderstood and to elicit a violent reaction than would an unexplained attack

The Masai pride began to stir toward dusk. The lions yawned cavernously, they stretched, and some ambled to the periphery of the rest area and defecated. Lionesses greeted one another, not casually but with a certain tautness that usually heralded a communal endeavor. Then abruptly they departed, heading for a nearby rise, ignoring the cubs, which of their own accord stayed behind. A broad valley stretched to the west and on it grazed scattered herds of wildebeest and zebra. Gathering at the rim of the valley, the lions surveyed the peaceful scene and waited while the plains changed from a delicate pink to a shadowy purple and finally grew dark. The lionesses fanned out and silently moved ahead. I waited with the two males, who stood motionless. Suddenly hoofs drummed the ground and then a wild scream shattered the night, the frenzied cry of a dying zebra. The male lions and I hurried toward the sound and by the time we reached the kill site the air was already heavy with the odors of blood and sour rumen contents. A writhing, tawny mass of lions covered the zebra, filling the night with their menacing growls, a drama of such naked emotion and terrible power that my rudimentary hackles raised themselves in subconscious apprehension. Within thirty minutes the zebra was dismembered and each animal crouched by itself, cleaning the meat off some bone. Brown Mane had taken possession of the whole rib cage and head.

It was now 8:20 p.m. Mrs. Notch headed toward the cubs, waiting a mile and a half away by the pool, and a few minutes later Flop-ear followed. All returned at 9:30 p.m., the youngsters trotting eagerly ahead of their mothers. While the cubs joined Brown Mane at the kill, the others rested nearby. After the last bones had been rasped clean, the cubs sought out their mothers and suckled in spite of being gorged with meat. At 1:20 a.m. the pride moved on, and again the cubs stayed behind, this time in some tall grass not far from the kill. Topping a ridge, the lions saw a herd of wildebeest. The lionesses spread out, only their dark heads visible in the moonlight as they waited for the prey to drift closer. For nearly two hours they sat, until finally a careless wildebeest stumbled into a hidden lioness. Within fifteen minutes the carcass was torn to pieces. This time Flop-ear departed first to fetch the cubs and in about ten minutes I heard her soft, hollow uh-uh signifying "come" as she led the youngsters to the kill. Black Mane provided meat for them. The first light of day found the pride resting once more, happily glutted. But the animals roused themselves to greet the sun with a communal concert of roars before once again sinking into contented oblivion. Most pride members had walked a mere 2.5 miles in twenty-four hours, they had twice eaten, and each cub had also suckled for an average of fifty-three minutes—lion life at its best.

I studied prides primarily during the dry time of the year. During the rains my attention was focused on the nomads of the plains. In that boundless space, the movement of the earth became almost palpable when I closed my eyes. Adrift in the universe, I felt a strange sense of loneliness, yet it somehow intensified my love for the area, and there on the plains I found a peace that no conscious thought could provide. When, as sometimes happened, another car intruded into my kingdom, I watched it with antagonism until it moved away. On moonlit nights the plains were at their most beautiful. Sometimes I left the car and walked up onto a rise and stood there, intoxicated by the wind and the stars far beyond. But a primordial fear would reassert itself, my imagination conjured up predators prowling the darkness, and I hurried back to the car, unable to escape my cultural bondage.

# LION

On the plains it was easy to follow lions and I did so often, sometimes only for a day or two, once for as long as twenty-one consecutive days, at such times aided by a radio attached to a collar around the animal's neck. Howard Baldwin, William Holz, and others took over some of these nightly vigils, but even so my enjoyment was replaced by crankiness after several days with little or no sleep. The main purpose of these intensive observations was to find out what lions do with their time, how far they travel, how much they eat, and so forth, knowledge needed to assess the impact of lions on the prey community. For example, we stayed with male No. 134 and his female companion, a middle-aged lioness, for nine consecutive days around the Gol Kopjes. The male was a handsome beast, about four and a half years old, with an open, unscarred face, and now in the first flush of maturity he attempted to establish a territory. One day's activity describes his routine:

A little after 6 p.m. many roars emanated from about half a mile beyond the kopje in which male No. 134 and his friend had been lying all day. They answered many of the calls, and at 7 p.m. they met males Nos. 103 and 104, probably brothers about four years old. The male and female charged and chased the two vigorously, at first bounding but then slowing to a trot for two miles, roaring repeatedly. After that strenuous interlude they either rested or wandered through several kopjes. Suddenly the male walked rapidly half a mile, where he met a young lioness chewing on a wildebeest head. She retreated about a hundred feet then fled when the male strutted in front of her. His

female arrived and gnawed on the head for several minutes. They visited several more kopjes, finally staying in one of them for their daytime rest.

The male had traveled 13 miles in 24 hours, most of them during the night; his average for the 9 days was 7½ miles. He ate on 7 of the nights—more often than is usual—three times on carcasses scavenged from hyenas and four times by joining lionesses. He lost his territory soon afterward. One morning I found him pacing restlessly near the Gol Kopjes without approaching his usual haunts. That night he abruptly turned and left the area when hearing roars there. On investigating, I found his old nemeses, males Nos. 103 and 104, in possession. Evicted from his territory, male No. 134 became fully nomadic again, still accompanied by his female friend.

For comparison, male No. 159, a nomad who had established himself around Naabi Hill one season, traveled an average of 6 miles a day, with a variation of 0 to 13.5 miles, during the 21 days that we followed him. He ate 7 times. Like many males, he seldom bothered to hunt for himself, except for an inept chase after a dik-dik that had ventured from the security of the thickets on Naabi Hill, and instead spent many hours each night just looking, listening, apparently hoping that food would somehow materialize. Once every few days he hiked several miles to a small pool where he drank. While lions will drink every day or two when water is available, this male twice waited for five days before making a reluctant trek to quench his thirst on the sixth.

One might assume that, with food everywhere around them, nomads on the plains move less than do other lions. But this is not so. The Masai and Seronera prides covered an average of only three miles a day on their hunts, and they worked no harder when their cubs were starving. Lions spend some twenty to twenty-one hours a day inactive—sleeping, dozing, sitting. To watch lions sleep in daytime is merely monotonous, but to keep awake at 2 a.m. with them still asleep is hard work, and sometimes I succumbed to the temptation to close my eyes. Almost invariably the lions chose that moment to vanish. Generally, lions spend less than an hour a day eating, but, of course, they are adapted to a feast-or-famine regime, during which several days may pass with no food, only to be followed by a truly gargantuan meal. An extreme example was male No. 10, an old scarred chap with a tatty yellow mane. He had scrapped with another lion

and been bitten severely in a leg. Seeking refuge in a kopje, he stayed there for seventeen days, barely moving, not eating, but probably subsisting on water from the small pools that had collected in the depressions of the rocks. Finally he ventured forth, healed but lean, and scavenged a carcass from hyenas. He gorged himself until his belly swayed grotesquely from side to side when he walked. I do not know how much he ate, perhaps almost one hundred pounds. Once I gave a male a weighed chunk of meat and in the course of a night he consumed seventy-three pounds, all there was, and he had not been particularly hungry when he started.

On the basis of what is known about energy metabolism in captive mammals, a lioness weighing 250 pounds would need about 7,300 calories per day and a male weighing 350 pounds about 10,300 calories, or 11 and 15½ pounds of food, respectively. Just how much lions actually eat can only be surmised. Male No. 134, for example, consumed an estimated twenty pounds of food a day during the period we followed him continuously, and male No. 159 about fourteen pounds. One group consisting of two males, three females, and eight small cubs killed four wildebeest in thirteen days, providing each adult fifteen pounds of meat per day, but a little of this went to the cubs. Sometimes lions ate more than they needed, sometimes no doubt less, so their actual consumption over a year was probably close to their requirement. A lion has to kill more than that, however, for up to 40 per cent of a carcass is waste, principally bones and rumen contents. Taking this into account, each female must kill or otherwise obtain about 6,000 pounds of prey a year and each male 8,400 pounds. A female could satisfy her needs with ten large zebra or 170 Thomson's gazelle. Naturally lions kill a variety of prey and on the average each probably takes about twenty to thirty animals a year.

To the lionesses falls the chore of catching most prey, as I described earlier. Not only do they have to support cubs and to some extent the males, but also incapacitated animals. The Young Female in the Masai pride, for instance, was bitten in the thigh during a tussle with the Seronera pride. Her hindleg withered until she was only able to hobble, ostensibly doomed to certain death, except that she eked out an existence at the

kills made by others until after about eight months she recovered. Not surprisingly, lionesses are excellent huntresses. They may be inept at times and they actually fail in their hunts more often than not, but that they can get by with working less than four hours a day speaks well for their ability.

Hunting lionesses have learned to take advantage of their environment. Darkness provides cover and at dusk they often wait near prey until their outlines blend into the surroundings. Thickets also provide concealment, enabling lions to hunt in daytime, and to see one of these cats creeping from bush to bush filled me with almost unbearable tension. I participated vicariously in the hunt: my muscles flexed as if for a rush, my breathing grew shallow, reverting to a predatory past when man too relied on the kill, until success or failure brought relief. "Of course the lion always approaches his prey against the wind," wrote one author, and similar statements have been made so often that they have achieved an aura of truth. Yet the Serengeti lions ignore the wind, even though they are much more successful in catching prey upwind than downwind. Herd size, the number of lions hunting together, and other factors also influence the outcome of hunts. The best way to convey the importance of some of these is to tabulate the success lions have in capturing prey once they have begun their stalk.

| | % HUNTS SUCCESSFUL |
|---|---|
| DAY HUNT | 21 |
| **NIGHT HUNT** | **33** |
| LITTLE OR NO COVER | 12 |
| **THICKET ALONG RIVER** | **41** |
| HUNT DOWNWIND (SEVERAL LIONS) | 15 |
| **HUNT UPWIND (SEVERAL LIONS)** | **48** |
| LIONESS HUNTING ALONE | 15 |
| **AT LEAST 2 LIONS HUNTING** | **30** |
| HUNTING SOLITARY WILDEBEEST | 47 |
| **HUNTING WILDEBEEST IN SMALL HERD** | **11** |

The ideal condition, the one with the greatest probability of success, would be for a group of lions to stalk a solitary large animal upwind near a thicket at night.

During a hunt, lionesses walk slowly, halting at intervals, looking and listening, often waiting for prey to wander into the vicinity, or, at night, to reveal its presence by braying or grunting. Some meals are served unexpectedly. One gazelle cautiously crossed a stream, its splashing awakening several lions as well as a crocodile. The lions crowded onto the embankment, watching the gazelle's progress, while the crocodile moved against the current almost wholly submerged, resembling a piece of bark with a will of its own. Sensing danger, the gazelle raced out of the water literally into the arms of the lions, and I can still hear the thud of the paw that knocked it down. One night the Masai pride discovered an occupied warthog burrow. While most members sat idly by, two lionesses tore at the sod with their paws until after an hour they had exposed seven feet of tunnel. Peering into the hole, one lioness slowly pulled a screaming pig from its retreat. Occasionally a lioness waits in ambush at a waterhole, sometimes for many hours, until finally some unsuspecting prey comes to drink.

But most hunts consist of stalks or, if there is a good opportunity to surprise some inattentive animal, a sudden sprint. I particularly remember a time when Flop-ear stalked across a scorched piece of terrain to catch a topi. An excerpt from my scientific report describes the hunt:

---

**At 1450 hours, a lioness sees three topi standing in the open near a river. After glancing at them briefly, she descends into a dry watercourse, moves downstream about 100 m,* then lies at the top of the embankment hidden behind a screen of grass where she watches a male topi standing alone on a burned area 50 m away and broadside to her. When at 1503 he turns away from her, she creeps into the open, partly hidden from him by a fallen tree. Some distant topi see the lioness and snort. The male jerks to attention and she freezes in mid-stride. But he fails to look around, and she suddenly trots forward, body held low, and is within 10 m of him before he spots her. He flees but she pursues and after 22 m grabs his rump with her paws, throws herself to the left, and both crash to the side, the topi's legs flailing the air. She lunges for his throat and bites it until he dies two minutes later.**

*1 meter (m) = 3.3 feet

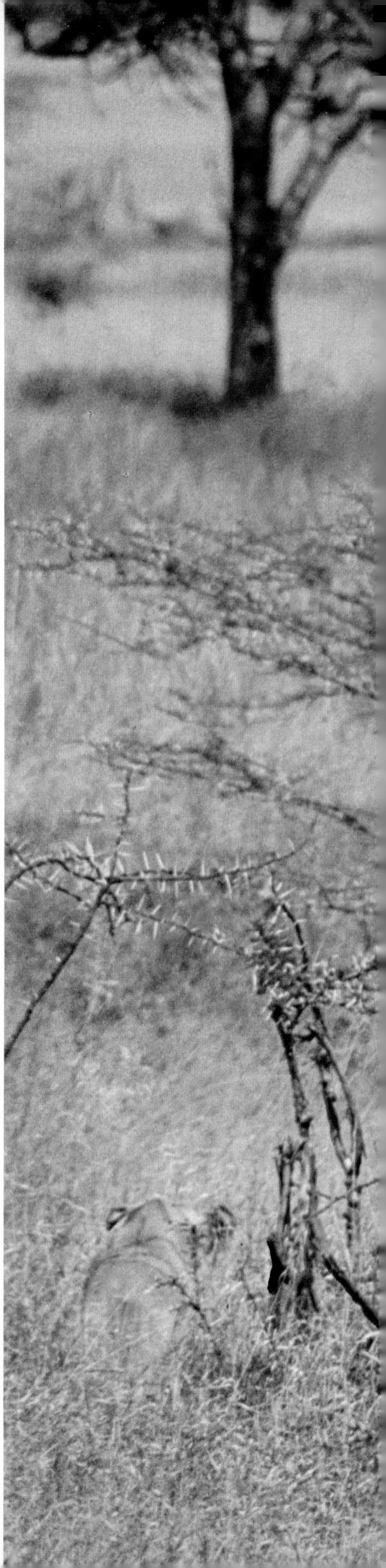

---

Much has been written about how lions kill prey. Small prey, such as Thomson's gazelle, present no problem, and they are merely grabbed with the paws or slapped down and dispatched with a bite in the neck. However, a large animal such as a wildebeest is torn down with such force, according to many authors, "as almost invariably to break the neck at once." This is seldom the case. Usually a lioness pulls her prey down after running up behind it, not by leaping at it, and then holds its throat, causing death by strangulation after several minutes. Occasionally a lion may place its mouth over the muzzle of a downed animal and in this way suffocate it. Tearing open the belly, lions then feed on the viscera, a fact that puzzles me, for intestines have a lower caloric content than the muscles. Perhaps they need the fat that is often deposited around the intestines, perhaps they obtain vitamins from the partially digested grass, some of which they invariably eat even though they have developed a special technique for eating intestine: they place one end of it on the tongue and lap it in slowly past the incisors, which help to squeeze out the vegetal contents.

But attacks do not always proceed smoothly. One day Flop-ear, the huntress who dispatched the topi so efficiently, was keeping her eyes on a river crossing used intermittently by zebra. A family of zebra crossed boldly but one adult tarried, uneasy about something. Flop-ear rushed and somehow bungled things, for she ended up clutching the zebra's neck with her forepaws and biting its nape while her body hung between its legs, exposed to the sharp hoofs. But the zebra merely spread its legs and stood rigid, screaming. Slowly Flop-ear pulled it down—right on top of herself. Attracted by the commotion, the rest of the pride ran up. Releasing her hold, Flop-ear gave the zebra a rare chance to escape, but it was overpowered after a short chase. The zebra family waited on the other bank, calling and calling, their only answer the growling of lions on the kill.

Before going to Africa I had read various accounts about the cooperative hunting strategies of lions. Rather skeptical at first, I was delighted that at least in some respects they are true, that lions are indeed so aware of the consequences of their actions both in relation to one another and to the prey that they have learned quite sophisticated techniques. Sighting prey, lionesses usually fan out and stalk closer until one is within striking distance. The startled herd may scatter or bolt to one side, right into a hidden lioness. A lucky hunt may yield three, four, even five animals. At times lionesses surround their quarry. While perhaps three crouch and wait, a fourth at one flank may backtrack and then circle far around and approach from the opposite side, a technique not unknown in human warfare. No obvious signals pass between the lions, other than that they watch one another, but years of association have no doubt taught them to coordinate actions without need for detailed instructions. A hunting technique may also be adapted to a particular situation. The Masai pride, for instance, often pursued prey at a spit of land formed by the confluence of two creeks. There several lionesses would sit and wait until gazelle entered the area, a cul-de-sac. Then they spread out and advanced on a broad front, quite in the open, having learned that the gazelle would not try to escape by entering the river thickets but would run back the way they had come. And there the lions waited.

Cooperation increases the lions' success not only in capturing prey but also in overcoming it. A lioness has no trouble pulling down an animal weighing twice as much as she does, but a buffalo, which may scale a ton, presents problems. One lioness and a young bull battled off and on for an hour and a half, the buffalo whirling around to face the cat with lowered horns whenever she came close, until finally she gave up and permitted him to walk away, only some scratches on the rump and a severed tail as remembrance of his encounter. But on another occasion five males came across an old bull. He stood in a swamp, belly deep in mud and water, safely facing his tormentors on the shore. Suddenly, inexplicably, he plodded toward them, intent it seemed on committing suicide. One lion grabbed his rump, another placed his paws on the bull's back and bit into the hump. Slowly, without trying to defend himself, the buffalo sank to his knees. There were no violent actions, no frantic movements, as the buffalo rolled onto his back and, with one lion holding his throat and another his muzzle, died of suffocation. It was a scene of such impersonal force that it achieved an elemental beauty.

The fact that lions with their relatively small brains and lack of a symbolic language use elaborate cooperative hunting techniques suggests that man too has been able to drive prey toward members lying in ambush, to chase prey into a cul-de-sac and so forth, since his beginnings as a hunter. Cooperation in capturing animals probably featured importantly in the evolution of human society, and it somehow strengthened my spiritual relations with the lions to know that we shared certain selective forces. At the same time it made me feel more distinctly human to realize that within my frame I carry both the terrible power of the lion and the frailty of the ape.

PRED

ATION

My study was too short to give a definitive answer to the question "What effect does predation have on the prey population?" in the Serengeti National Park. But some answer, no matter how tenuous, is needed and not on purely scientific grounds. Whereas man and his precursors were once threatened by nature, man now threatens nature instead. It is as stated in the Bhagavad-Gita: "Now I am become death, destroyer of worlds." With all hands turned against them, predators can hope to endure only in national parks and reserves. But even in these limited areas it is necessary to argue for their rights, to point out the economic benefits, the aesthetic reasons, the moral obligation for preserving them. We should not have to place a value on creatures to whom values are unknown, but should treat them with compassion and their world with restraint. Unfortunately that is not the way of man. Most parks are not self-sustaining ecological units in which a species can seek its own destiny, but arbitrary pieces of land often surrounded by cultivation. Some form of management may ultimately be necessary in all parks to keep the animal and plant communities reasonably well balanced and fit. Through eons of evolution predators have become excellent wildlife managers, far more discerning than man, and it behooves us for our sake and theirs to comprehend as well as we can their interrelationships with the prey.

As described earlier, adult lions in order to survive have to kill some 6,000 to 8,400 pounds of prey per year. Young animals need less and, taking this into account, the Serengeti lions require roughly 10,850,000 to 13,000,000 pounds of prey per year or some 40,000 to 72,000 animals. Hyenas need over 7,000,000 pounds, or at least 30,000 animals per year, according to Hans Kruuk, and the other predators also remove several thousand individuals. The total kill may be around 22,000,000 pounds, a substantial figure but not even 10 per cent of the amount available. Prey species can readily absorb a loss of this magnitude through reproduction. However, to present such a conclusion does little to clarify the subject of predation, for each species is affected differently. A key question that must be answered is to what extent predators take surplus individuals rather than healthy adults, whose loss depresses a population to a level below the one it would maintain if predation did not affect it. Since there is no space to discuss every species, and my information is scanty for many of them, I will merely give a few examples.

The age, sex, and health of prey that is killed depend in part on the hunting method used by the predator, especially in areas where several species use the same resource. Coursers such as the hyena and hunting dog select a specific individual and chase it, usually selecting the one that is easiest to catch. Consequently coursers kill a disproportionately large number of young and sick animals. While stalkers also capture such prey, they quite often depend on those that select against themselves by becoming vulnerable in some way, as by being inattentive, and because of this stalkers tend to kill more randomly from a population than do coursers.

Wildebeest are highly vulnerable during the first two months of life and they are killed by all large predators, particularly by hyenas, which may kill as many as 19,000 annually according to Hans Kruuk. But loss is meliorated in several ways. Calves are not only born at a specific season, making them unavailable for much of the year, but they also escape the many resident lions and leopards in the woodlands by spending the first few months on the plains. Although predators do kill many calves, their loss is so great—about 60,000 were estimated to have died early in life in 1966—that other factors must also contribute. Lost calves and those freshly dead from disease and malnutrition are commonly seen during the birth season, and one tragic day a herd of wildebeest forded Lake Lagaja, leaving about 685 dead calves in its wake.

After calves have reached an age of six months, only lion and hyena can readily kill them, and this they do quite selectively. Both predators capture proportionately twice as many bulls as cows. In a polygynous society, in which one male mates with many females, a large percent of the males are expendable members of society. Hyenas cull a large number of sick and old adults, too, those which are doomed to an early death anyway. On the other hand, lions predominantly kill animals in good condition. Taken as a whole, including individuals that are physically below par, lions harvested only about 3 per cent of the huge wildebeest population annually, and hyenas took perhaps a similar amount. With the wildebeest increasing at a potential rate of 10 to 15 per cent a year, the predators took about half, only a fraction of that consisting of healthy breeding stock. There were not even enough predators to weed out those in poor condition, and during the height of the dry season many mummified carcasses, untouched by predators, littered the woodlands, disease and malnutrition the cause of their lingering death. Between 1963 and 1970 the wildebeest increased from 322,000 to over 500,000, dramatic verification that the predators were unable to keep the population in check.

The impact of lion and hyena predation is even less when the breeding potential of the wildebeest is considered. In the early 1960's, when the population was fairly small, about a third of the cows had a calf in their second year. By the late 1960's, with the population high, only about 5 per cent had a calf in their second year. The wildebeest were controlling their own population and increased predation would simply have increased the birth rate.

The large and pugnacious buffalo has only the lion as a serious predator. About 6,300 yearling and adult buffalo die each year in the Serengeti, according to A. R. E. Sinclair, two-thirds of them through illness and malnutrition and one-third through lion predation, many of the latter being solitary bulls which in their old age have left the security of the herd to lead a misanthropic existence along river courses. The buffalo population has been increasing steadily, good proof that lions do not control it.

With wildebeest and buffalo escaping the full impact of predation—one by migrating, the other by being particularly large—the brunt of it falls on other species. Every tooth and claw seems to be turned against Thomson's gazelle, and it is not coincidental that females may have young twice a year. Gazelle are so difficult to count that I do not know whether the population is increasing, decreasing, or has remained stable. I suspect that the many predators have a greater impact on this species than on most others. While the migratory herds are on the plains, topi are the main prey species of the woodland lions. Reedbuck live primarily in river thickets, a favorite hunting ground of both leopard and lion. These and some other species have not increased markedly, as far as is known, and it seems likely that predation is a major factor limiting the size of their populations. Of course, predation works in conjunction with disease, accidents, and so forth, each contributing toward keeping a population in check.

Generally, when predation pressure is heavy, disease tends to be an unimportant cause of death and the reverse is true too. It is, therefore, significant that disease was more evident among wildebeest and buffalo than among topi, reedbuck, and other such woodland species.

While predation may be a major control on the size of a prey population, the ultimate factor is the habitat, the food supply. In the Serengeti the amount and distribution of the rainfall have a profound influence on the food supply. The vagaries of the seasonal rains may cause both drastic increases and decreases in the prey populations. The past years have been good and wildebeest and buffalo have responded to the seemingly unlimited food by increasing dramatically; the growth of other species appears to have been checked more by predators. Without rigid controls on their population, hoofed animals tend to increase up to and beyond the capacity of the range to support them. Finally, their food supply exhausted or unavailable because of a prolonged dry season, the animals die of starvation, which is often coupled with disease. Predators help to prevent such population explosions; they hold numbers below a level at which disease and starvation become important regulating factors. Only when predators are for some reason unable to exert themselves fully are sudden large oscillations in population size likely to occur.

The conclusion in my scientific report is also a suitable ending for this essay:

> The Serengeti predators are an integral and essential part of the ecological community. They help maintain an equilibrium in the prey populations within the limits imposed by the environment, they prevent severe fluctuations in the number of animals and condition of the habitat....The predators weed out the sick and old, they keep herds healthy and alert. The beauty of antelope, their fleetness and grace, their vital tension, are the evolutionary products of a constant predator pressure that has eliminated the stolid and slow. Man, one hopes, has gained enough wisdom from his past mistakes to realize that, to survive in all their vigor and abundance, the prey populations need the lion and other predators. Ecological and aesthetic considerations aside, predators should be allowed to survive in national parks without justification, solely for their own sake. Only by so doing can man atone in a small way for the avarice and prejudice with which he continues to exterminate predators throughout the world.

SELECTED READING: Although many books have been written about East Africa in recent years, I have selected only a few, which either provide detailed information about predators or best evoke the beauty of the area.

ADAMSON, GEORGE. *Bwana Game.* London: Collins; 1968.
Based on tame but free-living lions, this account provides fascinating glimpses into lion behavior.
ADAMSON, JOY. *Living Free.* London: Collins; 1961.
The most interesting volume of the famous Elsa trilogy.
AMES, EVELYN. *A Glimpse of Eden.* London: Collins; 1968.
A beautifully written narrative of a journey through East Africa's reserves.
EDEY, MAITLAND (photographs by John Dominis). *The Cats of Africa.* New York: Time-Life Books; 1968.
Well-written essays about each of Africa's cats, with excellent photographs.
GRZIMEK, BERNHARD, AND MICHAEL GRZIMEK. *Serengeti Shall Not Die.* London: Hamish Hamilton; 1960.
A classic account, which helped to preserve the Serengeti.
GUGGISBERG, C.A.W. *Simba.* Capetown: Howard Timmins; 1961.
A readable and useful summary of knowledge about lions up to 1960.
KRUUK, HANS. *The Spotted Hyena.* Chicago: University of Chicago Press; 1972.
A scientific report on the hyena in the Serengeti area.
SCHALLER, GEORGE. *The Serengeti Lion.* Chicago: University of Chicago Press; 1972.
A scientific report on the lion and other Serengeti predators.
VAN LAWICK-GOODALL, HUGO, AND JANE VAN LAWICK-GOODALL. *The Innocent Killers.* London: Collins; 1970.
Notes on the family life of the hunting dog, hyena, and jackal.

## IDENTIFICATION OF ILLUSTRATIONS

PHOTOGRAPHS BY THE AUTHOR

## A NOTE ABOUT THE AUTHOR

Born in 1933, George Schaller received his B.A. and B.S. degrees in zoology and anthropology from the University of Alaska and his Ph.D in zoology from the University of Wisconsin. He is a research zoologist at the New York Zoological Society and an adjunct assistant professor at Rockefeller University. Mr. Schaller's previous book for the general reader, *The Year of the Gorilla,* was published in 1964. In addition, Mr. Schaller is the author of scientific monographs giving the results of his field work: *The Mountain Gorilla, The Deer and the Tiger,* and *The Serengeti Lion.* When not abroad studying wildlife, he and his wife and two children live in Ryegate, Vermont.


## DESIGN & PRODUCTION NOTES

This book was set photographically by Composing Room+ Graphic Arts Typographers, Inc., New York, N.Y., and printed and bound by Arnoldo Mondadori Editore, Verona, Italy.

The text type, Stymie Light, and the display type, City Bold, are based on the Linotype face Memphis. Although both Stymie and City are derivative faces, they are more advanced styles that reflect greater simplicity.

Design & graphics were directed by R. D. Scudellari.

Production was directed by Ellen McNeilly.